The Classification of Sex

The
Classification
of
Sex

Alfred Kinsey
and the
Organization
of Knowledge

Donna J. Drucker

University of Pittsburgh Press

A portion of chapter 3 appeared in a different form as "'A Noble Experiment': The Marriage Course at Indiana University, 1938–1940," *Indiana Magazine of History* 103 (September 2007): 231–64. It is reprinted with the permission of the Trustees of Indiana University.

Portions of chapter 4 and 5 appeared in a different form as "Keying Desire: Alfred Kinsey's Use of Punched-Card Machines for Sex Research," *Journal of the History of Sexuality* 22 (January 2013): 105–25. It is reprinted with the permission of the University of Texas Press.

Published by the University of Pittsburgh Press, Pittsburgh, Pa., 15260
Copyright © 2014, University of Pittsburgh Press
All rights reserved
Manufactured in the United States of America
Printed on acid-free paper
10 9 8 7 6 5 4 3 2 1

Library of Congress Cataloging-in-Publication Data

Drucker, Donna J.
The Classification of Sex: Alfred Kinsey and the Organization of Knowledge / Donna J. Drucker.
    pages cm
ISBN 978-0-8229-6303-5 (paperback)
1. Kinsey, Alfred C. (Alfred Charles), 1894–1956. 2. Science—Methodology. 3. Classification of sciences. 4. Research—United States. 5. Sexology—United States. I. Title.
HQ18.32.K56D78 2014
306.7—dc23   2014012638

Contents

Acknowledgments
vii

Introduction
1

1. Learning the Trade, Creating a Collector
14

2. The Evolution of a Taxonomist
38

3. Teaching Life and Human Sciences
63

4. Ordering Human Sexuality
88

5. The Taxonomy and Classification of Human Sexuality
116

6. The Boundaries of Sexual Categorization
142

Conclusion
164

Notes
173

Bibliography
209

Index
235

Acknowledgments

THIS BOOK HAS its origins in two Indiana University (IU) graduate courses, one taught by Elin K. Jacob in the School of Library and Information Science (now the Department of Information and Library Science in the School of Informatics and Computing), and one taught by Judith A. Allen in the Department of Gender Studies. Jacob's course "The Organization and Representation of Knowledge and Information" sparked my abiding engagement with the historical and contemporary theories and practices of classification. Allen's spring 2003 course "Kinsey's Women: Genealogies and Legacies" traced the history of sex research related to women in conjunction with the fiftieth anniversary of the publication of *Sexual Behavior in the Human Female*. I sketched my earliest ideas on the relationship between Kinsey's entomology and sexuality research in the seminar paper for the course. Allen became my adviser and mentor, and as a result of her encouragement in my research on Kinsey, I took on the role of coordinating the Women's Sexualities: Historical, International, and Interdisciplinary Perspectives Conference held at IU in the fall of 2003. There I was introduced to an international and interdisciplinary group of scholars passionate about the study of sexuality, and I knew that I wanted to become a long-term part of that research community. My interests in the histories of classification, sexuality, and the life and human sciences have now coalesced into this book.

I thank Colin R. Johnson, Michael McGerr, Stephanie A. Sanders, and Eric Sandweiss at Indiana University for their insights and advice. Eric Sandweiss, as editor of the *Indiana Magazine of History,* helped me craft my first academic article on Kinsey's marriage course over good conversations at the *IMH* offices and Soma Coffee House. It appears as part of chapter three. Whether we were in Bloomington or Warsaw, talking in person or online, Colin R. Johnson has continually encouraged me to think ever more creatively about gender and sexuality. I am grateful to Peter Hegarty for his insights into gall wasps, statistics, and Lewis Terman's impact on Kinsey. Robin C. Henry deserves special thanks for helping me think through Kinsey's ideas about race. I had thoughtful discussions with Ronald Ladouceur about Kinsey's high school biology textbooks. Jared Richman provided critical feedback on the introduction and second chapter. Audra J. Wolfe provided high-quality guidance through the acquisitions process and excellent assistance with chapter six and the conclusion. For many good conversations I have had about Kinsey's life and work, I thank Howard H. Chiang, Rebecca L. Davis, Stephen Garton, Lynn K. Gorchov, Raymond J. Haberski, Jr., Robert E. Kohler, Agnieszka Kościańska, Carolyn Herbst Lewis, Joshua P. Levens, Donald W. Maxwell, Joanne Meyerowitz, Clark A. Pomerleau, Jeremy Rapport, Miriam G. Reumann, and David Serlin. Thanks are also due to the two anonymous reviewers for their comments and thoughts on an earlier version of the manuscript.

Librarians and archivists at the following institutions were indispensable to completing this book: American Museum of Natural History Library; American Philosophical Society Library; Archives of the Gray Herbarium at Harvard University; Arnold Arboretum at Harvard University; Bancroft Library at the University of California, Berkeley; Bentley Historical Library at the University of Michigan; Center for the Study of History and Memory at IU; Charles Babbage Institute at the University of Minnesota; Harvard University Archives; Indiana University-Purdue University Indianapolis Archives; Indiana University Archives; Kinsey Institute Library and Special Collections at IU; Lilly Library at IU; and Technische Universität Darmstadt. This book has benefited especially from years of assistance from Liana Zhou, Shawn C. Wilson, and Catherine Johnson-Roehr at the Kinsey Institute. Amberle Sherman and Alex Wolfe of the University of Pittsburgh Press skillfully guided this book through the publication process.

I am grateful to the following entities for financial support: Department of History, Indiana University; College of Arts and Sciences, Indiana University; Department of History, Indiana State University; Bentley Historical Library, University of Michigan; Library Resident Research Fellowship, American Philosophical Society; Department of Feminist and Gender Studies, Colorado College; Department of History, Colorado College; Technische Universität Darmstadt; and the Deutsche Forschungsgemeinschaft. Special thanks are due to Mikael Hård and Petra Gehring, the chairs of the Graduiertenkolleg Topologie der Technik at the Technische Universität Darmstadt, for accepting me as a postdoctoral fellow and thus providing me with the financial support and the time that I needed to finish this book.

I have had the good fortune to present parts of this research at conferences and workshops around the world and to benefit from perceptive audience questions and discussions. They include the Women's Sexualities: Historical, International, and Interdisciplinary Perspectives Conference in Bloomington, Indiana; the Annual Conference on Illinois History; the Biennial Women's and Gender Historians of the Midwest Conference; the American Historical Association; the Evolution and the Public Conference at Universität Siegen, Germany; Darwin Now: Darwin's Living Legacy Conference in Alexandria, Egypt; the Southern Association for the History of Medicine and Science Conference; the U.S. Intellectual History Conference; the Archival Technologies and the Genealogy of Datapower Workshop at the Max Planck Institute for the History of Science in Berlin, Germany; and the Techno-Topologies: Spatial Perspectives-Spatial Practices Conference at Technische Universität Darmstadt.

Acknowledgments **ix**

The friends and family members who have supported me and my research include Dana Umscheid Autry, John Philipp Baesler, Tobias Boll, Erin Grip Brown, Catarina Caetano da Rosa, Laura Clapper, Kathleen A. Costello, Fr. Dave Denny, Pat DeVita, Fr. Daniel Donohoo, Adrienne Drucker, Alan J. Drucker, Joey Drucker, Amy L. Elson, Regula Meyer Evitt, Rachel E. Feder, Kathi Fox, Stefan Glatzl, the late Peter A. Kraemer, Deborah A. Kraus, Anne Lucke, Jory Drucker Mangurten, Kris Pangburn, Charles Peters, Sonja Petersen, Mark A. Price, Bahar Şen, Jennifer Stinson, Betty Watson, Charles Watson, Katie Watson, Charles T. Wolfe, and Kevin Wong. My young cousins, Halle Mangurten, T.J. Mangurten, Shay Drucker, and Sage Drucker, bring me much joy. My parents, Donald S. Drucker and Diane K. Drucker, deserve special thanks for their love.

I have used only the initials of some of Alfred Kinsey's correspondents' names per request of the Kinsey Institute Archives.

The Classification of Sex

Introduction

> Most of us like to collect things. If your collection is larger, even a shade larger, than any other like it in the world, that greatly increases your happiness. It shows how complete a work you can accomplish, in what good order you can arrange the specimens, with what surpassing wisdom you can exhibit them, [and] with what authority you can speak on your subject. Taxonomists aren't so different from the rest of you who do a little collecting.
>
> —Alfred C. Kinsey, *An Introduction to Biology*

ALFRED C. KINSEY loved to collect, to study, and to classify elements of the natural world, and his enthusiasm for those scientific practices shaped his whole academic life. He shared his passion for collecting with the young readers of *An Introduction to Biology*, his first textbook for American high school students. For him, collecting led to happiness, but a larger collection led to even greater happiness. Collecting was a means of developing good character and showing scientific accomplishments. Having a collection of natural objects demonstrated classificatory abilities and handicraft skills, and provided an opportunity to teach others about one's area of expertise. As a professor at Indiana University (IU) in Bloomington, he wanted to teach high school and college-aged students, and anyone else who might be reading, that they all had the potential to use taxonomy to better understand the world around them. The quotation in the epigraph to this chapter summarizes how much collecting, categorizing, and analyzing facets of the natural world meant to him as a teacher, a scholar, and a writer. The division of the natural world was not merely essential to his scientific research but also was emblematic of his desire to make order in the world through classification.

The intention of this book is to examine the development of and patterns in Kinsey's research from his earliest work on gall wasps in the late 1910s through his *Sexual Behavior in the Human Male*, published in 1948 (hereafter *Male* volume) and his final major collaborative work, *Sexual Behavior in the Human Female*, published in 1953 (hereafter *Female* volume).[1] The link between his earliest work and his latest work is his focus on methods of mass data gathering, classification of that data, and the knowledge that classification can generate. Kinsey's classification of insect data gave rise to the identification of new species in the genus *Cynips* and the rethinking and reordering of existing species. His classification of edible wild plants in eastern North America helped nature lovers identify and enjoy the fruits of the land. His classification of human sexual behavior data—up to 521 data points per interviewee, and often more information handwritten on each of the eighteen thousand sex history data forms—led to the publication of the most influential texts on human sexuality in the twentieth century.

Kinsey's commitment "to investigate honestly, to observe and to record without prejudice" and "to observe persistently and sufficiently" affected all of his scientific work.[2] His classification and reclassification of *Cynips* led to speciation of the genus that still stands in the present. His and Merritt Lyndon Fernald's work on edible wild plants remains a classic nearly one hundred years after they wrote it.[3] Kinsey developed a classification system for the Institute for Sex Research's art collection, and he and the Institute's first librarian developed a modified Dewey Decimal Classification system to catalog the ISR's library collections.[4] His classification of sexual behavior data led to the creation of the 0–6 (heterosexuality–homosexuality) scale, became a source of identification and community for gay and lesbian rights activists, became a source for changes in sex offender laws, debunked the myth of the vaginal orgasm, and provided little support for a theory of male–female sexual difference, among many other short- and long-term effects.

Kinsey's decision to move from studying gall wasps to human sexuality, to move from the life sciences to the human sciences, has intrigued many scholars, filmmakers, novelists, and the wider public, and this book shows that the connection between his two major, seemingly disparate fields is the gathering, organization, and classification of scientific data. Previous speculations on his reasoning include boredom, a desire for wider fame and renown, ambition to be "a second Darwin," or a yearning to understand or to justify his bisexual or homosexual desires.[5] Kinsey never wrote a reflection of his own motivations for the shift, which might have included any or all of those reasons in different combinations. The historical record makes clear that his research shift took place in larger scientific contexts: a change in the purpose of field collecting of specimens in the mid-1930s, subsequent changes in the use of laboratory animals for evolutionary studies, and the modern (a.k.a. evolutionary) synthesis.

Further, his discovery of vast new sets of data from the natural world of human sexual behavior excited and energized a field naturalist like himself. Also, his extensive work on sex education for his high school textbooks, high school workbooks, and Indiana state teaching standards led him to believe that the field of human sexuality studies would benefit from mass data collected and analyzed from a taxonomic perspective. A commitment to classifying the natural world structured the entirety of Kinsey's academic life, and that commitment was evident in each field he studied.

The title of this book, *The Classification of Sex: Alfred Kinsey and the Organization of Knowledge*, is a deliberate echo of the Case Western Reserve librarian Jesse H. Shera's book *Libraries and the Organization of Knowledge*, published in 1965.[6] Shera, at the time he was writing, faced an exponential increase in the amount of information production on the governmental level. He wanted to provide a guide for librarians seeking to manage an unprecedented level of materials in all sorts of new media, in order to catalog and to merge them with older forms of media. At the time, Shera argued that librarians needed to be flexible regarding the many different ways that they could classify ever more complex bodies of data with ever more complicated forms of media. Librarians needed to be aware that their organizational skills were increasingly in demand, and that data users needed classification systems to be clear, up to date, and provide information in combinations that they may not have thought of before. Classification, in other words, was a tool to order information, without which researchers could neither find nor create new knowledge, and the ability to do it well was more important than ever.

Shera wrote in a time of information expansion, when librarians across the United States were seeking ways to manage masses of documents on both practical and philosophical levels for themselves and for library users. Kinsey's academic life involved a broad variety of organizational and information management skills critical to his professional success, and he developed those skills over time to manage bodies of continually expanding information about gall wasps, wild plants, and human sexual behavior. Such skills were especially important as he sought to understand patterns in living data that were not and never could be entirely fixed. All of the different tasks that Shera identifies for librarians—acquisition, identification, classification, comparison, and critical analysis—Kinsey took on in order to make his ever-shifting bodies of data intelligible for himself, and to make his arguments about that data convincing to his readers. This book demonstrates how Kinsey, with help from assistants and staff, completed the multitude of tasks necessary to gathering and ordering information so that he could make some sense out of the natural world and could produce new scientific knowledge.

Kinsey made his transition from studying gall wasps to human sexuality in the 1920s and 1930s, as human and social scientists in the United States were shifting toward quantitative analysis in order to supplement their qualitative studies.[7] The language of sociologists and other social scientists resembled the language that Kinsey would use in the 1940s and 1950s to assert the scientific, objective nature of his own work. For example, W. Lloyd Warner and Paul S. Lunt, authors of the first book in the "Yankee City" series in American sociology, used language to describe their social science work as part of "modern science" in ways similar to that of Kinsey in his approach to sex research. As they wrote in *The Social Life of a Modern Community*, "The three characteristic activities of modern science are the observation of 'relevant' phenomena, the arrangement of the facts collected by such observation into classes and orders, and the explanation of the ordering and classification of the collected data by means of so-called laws or principles."[8] Kinsey would likewise regularly describe his method of gathering sex history data by repeating the word "observation." Pitirim Sorokin, a prominent sociologist who became an opponent of Kinsey's after the *Female* volume was published, proclaimed in the 1920s that "the task of any scientific study is to define the interrelations of the studied phenomena as they exist."[9] Kinsey adopted similar terms to describe his objectivity and distance from the many sexual science researchers of the late nineteenth-century, such as Richard von Krafft-Ebing and Sigmund Freud, who made sweeping social or scientific judgments based on detailed descriptions of small numbers of cases.[10]

Furthermore, multiple intellectual shifts were occurring in the life sciences at the same time: "from inventory and classifying to research on the micromechanics of speciation in local populations"; from a morphological species concept to a biological and populational one; from completing inventories of specimens across species and taxonomic housekeeping to researching evolutionary biology in laboratories; from collecting specimens in the field to creating laboratory specimens far from it; and from delineating species and the relations between them to the process of speciation within single species.[11] Those changes occurred across the American life sciences beginning in the mid-1930s, combining simultaneous changes in experimental design, theoretical frameworks, and underlying assumptions about the relationship of animal behavior and evolution. In a short time frame, then, "the study of evolution in the United States [had] shifted from mapping the evolution of particular traits or behaviors to studying mechanisms of the evolutionary process."[12] The specific historical context of Kinsey's transition from gall wasps to sex research shows how new conceptual frameworks, as exemplified in the broad changes in practice effected by the methodological and theoretical changes of the evolutionary synthesis, can lead to a re-evaluation of established knowledge of the natural world. In Kinsey's case, the broader shift in life sciences from species discovery to speciation inspired him to take his taxonomic framework, information management abilities, and scientific method away from entomology and to apply it to a new field. The rejection of his evolutionary ideas pushed him out of entomology, and the lure of new raw data pulled him into sexology.[13] As taxonomies can be read "for the social orders that they read onto the materiality of life and the resulting actions that they legitimate," his taxonomies of human sexual behavior, particularly as embedded in the 0–6 scale, concretized his vision of how American society and individuals should order sexual life.[14] Kinsey's ability to gather, organize, classify, and analyze mass amounts of data in diverse fields in order to create new scientific knowledge was the hallmark of his academic life.

Kinsey's career in classification across two different areas of science reflects the history of the organization of knowledge in the mid-twentieth century. His classification methods produced quantitative knowledge about living (or formerly living) objects at a time in academia when scholars in life and human sciences were moving from qualitative analysis of individual specimens or small groups to large-scale quantitative analysis using machines. As "any system of knowledge . . . relies on robust, enduring techniques, technologies (even simple ones), practices, and recording methods," such techniques and technologies were adaptable across study objects in Kinsey's academic life. Furthermore, close examination of Kinsey's work practices mirrors the aim of the recent material turn in science studies: "To refocus attention on historical processes (social, technological) whereby information is produced by particular actors, [and] encoded on specific technologies that allow them to be stored and relayed over space and time."[15] Kinsey's use of material culture for ordering and storage, from Schmitt boxes to interview sheets to punched cards, shows how a single scientist can use and manipulate different media across a decades-long research career in the service of an overarching goal—to organize, to make sense of, and to share scientific knowledge with anyone interested in reading and discussing it. Without the development and use of specific technologies of recording, storage, and mechanization, it would not have been possible for Kinsey to produce his texts on gall wasps and human sexual behavior in their final forms.

Finally, Kinsey's academic life, through his efforts in evolutionary and human sexuality studies, is part of the history of the organization of knowledge. He aimed, to the extent possible, to grasp the processes and patterns that made up behavior in the organic human world. Early modern and modern European historians in particular have written about how scholars struggled to systematically organize their books, letters, notes, notebooks and other epistemic objects to find them easily when reading and writing. Noel Malcolm wrote of the mid-seventeenth century, "The project of gathering together and systematizing all existing knowledge seemed an absolute necessary first step towards the improvement, or even perfection, of human understanding."[16] Kinsey knew, by the time he completed *Sexual Behavior in the Human Female*, that even his thorough and systematic study of sexual behavior had just scratched the surface of what scientists might eventually find on the subject. Nonetheless, the *Male* and *Female* volumes together represent the determination of one man, plus one multi-person research institute and numerous "friends of the research," to manifest comprehensive understanding of human sexuality in print.[17]

Historiography

Scholars have scrutinized Kinsey and his work from multiple perspectives, including history and biography, sociology, gender/sexuality studies, and the history of science. Few of them investigate his reading and scholarly development, and fewer still investigate his classification practices. Kinsey has been the subject of four book-length biographies, two written by Institute for Sex Research/Kinsey Institute (ISR/KI) staff members (Cornelia V. Christenson [1903–1993] and Wardell B. Pomeroy [1913–2001]), one by an academic historian (James H. Jones), and one by an independent author (Jonathan Gathorne-Hardy).[18] Pomeroy wrote a conversational biography of Kinsey, which focuses more on the personalities and social lives of staff, friends, and visitors of the Institute than its research, and includes little analysis of the *Female* volume.[19] Christenson, who joined the staff in the early 1950s, also passed over the Institute's research without much depth. Jones's biography of Kinsey, *Alfred C. Kinsey: A Public/Private Life*, was the first to link Kinsey's personal life directly and negatively to his research.[20] Jones, like Pomeroy and Christenson, was concerned more with the personalities and sex lives of the ISR/KI staff than with the actual research that they and Kinsey conducted.[21] Gathorne-Hardy covers much of the same terrain as Jones's *Alfred C. Kinsey*, albeit in a more positive vein.[22]

Other historians have studied Kinsey's work as part of broader sociocultural analyses of sex, gender, statistical methods, and surveys, including Lynn K. Gorchov, Sarah E. Igo, Elaine Tyler May, Regina Markell Morantz, and Miriam G. Reumann.[23] Kinsey's work also has a significant place in three broad surveys of the history of sexuality, including *Intimate Matters: A History of Sexuality in America*, *Twentieth-Century Sexuality: A History*, and *Histories of Sexuality: Antiquity to Sexual Revolution*.[24] Jane Gerhard situates Kinsey in the history of American second-wave feminism, Jennifer Terry examines his importance to the history of homosexuality in the United States, and Paul Robinson places him more specifically between Havelock Ellis and William H. Masters and Virginia E. Johnson in an intellectual history context.[25] The sociologists Julia A. Ericksen and Janice M. Irvine place Kinsey in two overlapping yet narrower contexts: his position in the history of sex surveys and his location in the history of sexology.[26] Kinsey also figures into histories of American biological and animal sciences, including Joshua P. Levens's "Sex, Neurosis, and Animal

Behavior," Philip J. Pauly's *Biologists and the Promise of American Life: From Meriwether Lewis to Alfred Kinsey*, and Robert E. Kohler's *Landscapes and Labscapes: Exploring the Lab-Field Border in Biology* and *All Creatures: Naturalists, Collectors, and Biodiversity, 1850–1950*.[27]

This book traces the ways that classification shaped Kinsey's academic life as he moved from one project to another and often was involved in several pursuits at the same time. It shows how Kinsey's patterns of studying, teaching, and synthesizing large amounts of quantitative and qualitative information structured his career as a professor and researcher over forty years of activity. It connects different areas of Kinsey's scholarship to establish how they influenced each other, such as the effect his research on edible wild plants had on his thinking about racial difference. It highlights the ways that Kinsey made classification of bodies, bodily processes, and behaviors foundational in the new field of sexology, incorporated machines into sex research, and centered behavior, not just identity, as a field of critical inquiry in sexology. Lastly, it supports Colin R. Johnson's contention that "gender and sexuality are not just socially constructed, they are historiographically constructed."[28] The Kinsey Reports establish that Kinsey's thinking on human gender, though "gender" was not yet an intellectual construct, blended past scholarship and his own data and ideas to form new ways of thinking about the contours and meanings of manhood and womanhood.

Kinsey, Classification, and the History of Science

This book argues three main points about Kinsey and his work. First, day-to-day work practices at each stage of Kinsey's research affected his final work products, and detailed analysis of those practices leads to a more in-depth understanding of the processes of creating knowledge in the life and human sciences. As Bruno Latour wrote in his classic text *Science in Action*, "The history of a science is that of the many clever means to transform whatever people do, sell and buy into something that can be mobilised, gathered, archived, coded, recalculated, and displayed."[29] Latour frames the question of what makes a science as a question of practices of collecting, defining, ordering, and analyzing data. Studying Kinsey's work practices, from his earliest gall wasp research to the publication of *Sexual Behavior in the Human Female*, shows how he, and later his assistants, transformed mass collections of entities, from gall wasp wing vein formations to instances of orgasm via masturbation, into data suitable for scientific analysis. Paul Robinson was incorrect to say that for Kinsey "taxonomy was intended more as a critical than as a constructive tool."[30] Clearly, Kinsey used his taxonomic skill to create new forms of classification, such as the 0–6 scale.

Kinsey's lifelong pattern of collecting centered on gathering large amounts of raw data before he organized that data into categories. After data organization, he was then able to assemble the insect data into descriptions of new species or revised descriptions of older species, and sex history data into numerical data via punched-card machine. He coupled quantitative data with qualitative data to produce extensive portrayals of insect species and of human sexual behavior. Studying Kinsey's work practices—which he in turn taught to his assistants and fellow ISR staffers as his projects matured—sheds light on the creation of the *Male* and *Female* volumes, their similarities and differences, and how they led him to consider a universal theory of human sexuality.

As Kinsey established practices and techniques for collating his data and organizing it into patterns, he was then able to form connections between bodies of data to postulate new species of gall wasps and to establish connections between types of sexual behavior. He embraced the available technologies for data manipulation. The process of using the many different configurations available in punched-card machines, for example—"the physical technologizing of knowledge"—influenced how well Kinsey was able to put together the different pieces of the human sexuality puzzle that he had at hand. According to Latour, "The history of technoscience is in large part the history of all the little inventions made along the networks to accelerate the mobility of traces, or to enhance their faithfulness, combination, and cohesion, so as to make action at a distance possible."[31] The puzzle pieces that Kinsey had—items of sexual history data from diverse individuals and sources around the United States, Canada, and beyond—were initially "at a distance" from each other, but his research practices and data analysis techniques made it possible for him to make them cohere into a clearer picture of the whole of human sexuality.

Secondly, close analysis of Kinsey's work demonstrates the contentions of historians, of the historian of science Robert E. Kohler, and of library science scholars that methods of data gathering and classification are historically contingent: that they take place depending on scholars' existing knowledge of the data, how much others have already collected, how much others have already classified, and the strength of previous researchers' extant analysis.[32] For Kinsey, his organization and classification techniques were both built on the facts that he was gathering data that few others had tried to assemble in any depth, and that he had the freedom to organize and to classify that new, raw data as he saw fit. As a doctoral student, Kinsey trained as a taxonomist of insects, and that training fostered in him the ability to discern order in mass quantities of data, whether insect or human, in such a way that would reveal new patterns and connections between cataloged and previously uncataloged data. Most life and human scientists, past and present, develop some skills in classification, or else the analysis of individual specimens and the classification of species would be overwhelming and nearly impossible. However, Kinsey made the classification of data into its own art form, using equal amounts of care and precision in detailing the intricacies of cells on a gall wasp's wing and in creating an accumulative incidence curve for subjects' first instance of a particular sexual behavior. As Shera puts it, "Classification is the crystallization or formalizing of inferential thinking, born of sensory perception, conditioned by the operation of the human brain, and shaped by human experience. It lies at the foundation of all thought, but it is pragmatic and it is instrumental."[33] Kinsey's intuitive ability to classify masses of data with great precision was the foundation of his lifelong scientific practice.

Shera's observations about the importance of classification to librarians clearly apply to Kinsey, the inveterate taxonomist: "He must appreciate classification, not as a tool, but as a discipline in which is to be studied the reaction and response of a living mind to the record left by a distant and usually unknown mind; a discipline that seeks to achieve a better understanding of the changing patterns of thought and the points of contact at which they can be related to specific units of recorded information."[34] The insect world and the world of human sexual behavior provided much insight into "changing patterns of thought" as they manifested in human society, or more obscurely through evolution. The insects, along with records of human sexual experience, provided Kinsey with "specific units of recorded information" that he could then order to provide new insights and to create new knowledge. Readers, whether they had a professional interest in the *Male* and *Female* volumes or not, could interpret and use the knowledge Kinsey presented to inform their personal or professional experiences as they saw fit. Many gay and lesbian readers in particular would cite the Kinsey volumes as inspirations for their social action and organizing from the late 1950s forward. As Anne Fausto-Sterling states in *Sexing the Body*, "With the very act of measuring, scientists can change the social reality they set out to quantify."[35]

Kinsey's appreciation for the explanatory potentials of classification led him to ordering data into horizontal scales, charts, and graphs for ease of analysis and interpretation by others. Other sociology and biology researchers were similarly using scales to demonstrate the ranges of variation in their work.[36] While many researchers ordered their data on a single linear scale, the complexity of Kinsey's data led him to order his large quantities of data, particularly his human sexual behavior data, into multiple scales. He then correlated those scales with each other to reveal linkages and patterns between different aspects of behavior and between biological or social characteristics and behavior. Both narrowly conceived scales and scales aggregating hundreds of thousands of data points were necessary to display and to explain data adequately. According to two scholars of ontology, "A multiplicity of ontologies—of partial category systems—is needed in order to encompass the various aspects of reality represented in diverse areas of scientific research." Sometimes Kinsey argued for cause-and-effect relationships between groups of data, as when he suggested that premarital sexual experience to orgasm helped women to be more orgasmic after marriage.[37] Some critics, however, argued that Kinsey saw causation where they saw only correlation between types of behavior and social characteristics. In any event, however, even multiple scales and Kinsey's detailed explanations of the intersectionality of data failed to capture the full diversity of human sexual behavior that Kinsey had found in his investigations. To his own frustration, he was unable to articulate a synthetic theory of human sexuality across all aspects of body and mind.

Third, sexual science can never be "objective" in such a way as to satisfy all possible critiques. Ideas of objectivity and the research practices that enact those ideas are historically contingent, and "first and foremost, objectivity is the suppression of some aspect of the self, the countering of subjectivity."[38] Individual observers' ideas of what objectivity means varies with each observer and his/her background, training, and interests. Kinsey's work could never be objective enough to answer the criticisms of everyone, past and present, concerned with the "truth" or "reality" of sexuality. Kinsey knew that his work could never be satisfactory to all readers, and also that readers would not separate him from his data and publications. He says as much in the *Female* volume, when discussing the possibility of gender bias in his research: "It would be surprising if we, the present investigators, should have wholly freed ourselves from such century-old biases and succeeded in comparing the two sexes with the complete objectivity which is possible in areas of science that are of less direct import in human affairs. We have, however, tried to accumulate the data with a minimum of pre-judgment, and attempted to make interpretations which would fit those data."[39] Thus sexual science, like other sciences, could not be universally objective to all of its practitioners and readers. Like any scientific research, it had its own kinds of limitations, but those limitations become particularly vivid given the wide interest in understanding and explaining the sexual behaviors, desires, identities, and interests of humans. Readers tend to have high expectations for scientific research on human sexuality, and are often disappointed when their expectations are not met. Understanding the limits of scientific objectivity, particularly when it comes to human sex research, helps readers consider such research within the researcher's own framework, and limits their expectations that any one researcher will develop a universal theory of human sexuality that explains all behavior across space and time.

Chapter Overview

Chapter one describes the establishment of Kinsey's research methodology through his graduate study and study of gall wasps. Kinsey trained as an entomologist under William Morton Wheeler at Harvard University's Bussey Institution. Wheeler modeled an ideal type of scientist that Kinsey later emulated: one active in professional organizations and building professional relationships, one who linked masculinity with field gathering, and one who also had a strong interest in the connections between entomological and human studies. Kinsey developed a standard taxonomic practice centered on gathering large and comprehensive amounts of specimens, trust in naked-eye observation above manipulative laboratory techniques, and an interest in discovering and describing new species. As Kinsey contemplated a transition to sex research in the mid- to late 1930s, the parameters of taxonomic research began to shift.

Chapter two investigates how Kinsey's interest in evolution and in sex research developed over the course of writing a guidebook to edible wild plants, three high school textbooks, workbooks, and a life science teaching guide with evolutionary content and also through teaching sex education. Kinsey began his transition from horticultural and entomological to sexological research at the moment when his fellow biologists were shifting from mapping traits or behaviors to studying the mechanisms of the evolutionary process during the evolutionary synthesis. By making that shift, Kinsey became a contributor to the nascent, lively, and growing academic field of sexology that was in transition from a case history model to a numerical and mathematical model. Rather than shifting to studying the process of evolution in gall wasps—an animal that reproduced infrequently and was difficult to breed artificially—Kinsey shifted to studying the process of how humans developed as sexual beings. Kinsey's move from entomology to sexology also took place at a historical moment when biologists were leaving behind survey collecting in order to name species and were embracing targeted collecting as an instrument of evolutionary theory. His shift from entomology to sexology from the late 1930s onward allowed him to maintain his methodology and collecting patterns and to continue the fieldwork that stimulated his intellectual energies.

While chapter two examines Kinsey's research and writing of academic treatises and high school textbooks as a means of understanding how he chose to depart the worlds of entomology and evolution, chapter three focuses on the teaching topics that pulled him into the world of sex research. Kinsey's outlines and lecture notes for his biology and evolution courses from the late 1920s through the early 1940s capture his interest in teaching students to think, to question their beliefs, and to make decisions for themselves. The chapter then shifts to the research behind and the history of the interdisciplinary IU marriage course. Examining the marriage course and Kinsey's early sexological field work in Chicago and northern Indiana illustrates how Kinsey developed the first sex history interview as a result of establishing working relationships with university students, faculty, and members of homosexual and gay-friendly communities. The early sex history interviews revealed a potential, almost limitless body of data on human behavior that captured Kinsey's interest, energy, and imagination.

Chapter four centers on the techniques that Kinsey used to develop his sex history interview method, the interview recording sheet, the process of recording interviewee responses, and the statistical methods used to analyze the recorded data. It also examines how he transformed the handwritten data into numerical data via punched-card machine. He based his recording scheme on a similar one that he had developed for organizing gall wasp morphological data. In the fall of 1939, he had several meetings with the biometrician Raymond Pearl, and with him decided on statistical and sampling methods that would suit the project and would provide the best possible representation of a broad range of Americans given personnel and time limitations. After the IU marriage course concluded in September 1940, he began renting punched-card machines, and he and his staff transferred the data from the interview grid to the punched-card grid, enabling complex statistical manipulations of vast data sets. Additionally, data sorted via punched-card machine allowed Kinsey to form data into horizontal scales, which displayed complex data sets in easily readable and understandable forms. Kinsey's patterns of data classification, organization, and manipulation enabled the production of the *Male* and the *Female* volumes and their startling conclusions about the diversity of human sexual behavior. By downplaying qualitative narrative methodology in favor of quantitative data-gathering, Kinsey set aside the form of data collection that was gradually losing favor in humanist fields. His shift to using machine-organized data and quantitative methodology signaled a shift in the most popular and dependable tools used in the human sciences.

In the *Male* volume, Kinsey describes and defends his taxonomic technique and statistical methodology as the best means of obtaining and analyzing human sexual behavior data. His techniques led to elements such as the 0–6 scale, which characterized amounts of same- and opposite-sex behaviors in an individual. Kinsey also used the punched-card data to place educational and social class data on scales to consider in the relationships between age, education, social class, and numerous other personal attributes. A three-man American Statistical Association (ASA) review team questioned Kinsey's statistical technique.[40] In the *Female* volume, following the advice of the ASA statisticians, Kinsey would downplay the relationship between social class and sexual behavior, eliminate data from women in prison from his sample, and not write chapters on sources of sexual outlet in women. The discovery of extensive variation in sexual behavior in the *Male* volume led Kinsey to study the relative effects of psychology, anatomy, physiology, hormones, and the brain on human sexual behavior; to create the most comprehensive synthesis of human sex research that he and his team could manage; and to find possible reasons for male–female sexual difference in the *Female* volume.

As Kinsey researched and wrote the *Female* volume, he brought many new voices and much new information into the process. New contributors and information from new sources, including criminologists, lawyers, obstetrician/gynecologists, animal behaviorists, and liberal religious leaders, along with the Institute's own animal and human behavior filming, broadened Kinsey's research to encompass his effort to theorize human sexuality wholly. He was left with conflicting data regarding reasons for the similarities and differences in men's and women's sexualities. His decision to focus on psychological "conditionability" as an explanation for men's and women's sexual differences led him to downplay sociocultural reasons. Though Kinsey was not able to advance a synthetic theory of sexuality, examining the process of the *Female* volume's creation reveals the broader systematic processes of sexual knowledge acquisition and organization via Kinsey's deployment of the experimental systems of the life and human sciences.

A strong sense of the importance of mass collection and naked-eye observation undergirded Kinsey's gathering, description, and ordering patterns from his earliest gall wasp work through his wild plant work and his human sex research. Kinsey's gall wasp, wild plant, and human sexual data mirror each other through his emphasis on the detailed labeling and recording of each data object, the maintenance of flexibility for the manipulation of each object, and his prioritizing of mass yet targeted collecting. Through a focus on the precise material constitution of his research processes, Kinsey was able to produce publications based on large data sets that he believed could in turn support broad conclusions about the evolution of species and human sexual behavior if he (and his graduate students or coworkers) carefully managed, controlled, and analyzed them all. Historicizing Kinsey's and the Institute for Sex Research's collection practices and management as well as their scientific knowledge production unearths the epistemic webs that link them together and reveals the historical processes of structural changes in these scientific systems of knowledge. Scrutinizing Kinsey's research methods shows how a scientist's intense focus and emphasis on naked-eye observational techniques and practices can configure an entire career even through a seemingly dramatic shift in study object. It also reinforces the idea that "pure" scientific objectivity is nearly impossible to achieve. Regardless of study subject, some unknown elements and mysteries slip through even the most highly regulated scientific research practices.

I

Learning the Trade, Creating a Collector

I must have passed on to you [William Morton] Wheeler's advice, that the research I did in the first ten years out would determine my entire future in research.

—Alfred C. Kinsey to Ralph Voris, March 26, 1931

If a student could do research at the Bussey he could do it anywhere.

—Karl Sax, "The Bussey Institution"

ALFRED KINSEY HAD a strong curiosity about the life sciences from an early age—an enthusiasm that guided the course of his life and career. He wrote in his second high school biology textbook in 1933 that his inspiration for writing textbooks was "in the experiences of thirteen summers in camps where I had a share in showing plants and animals to some ten thousand young folks of high school age." He archived the work of boys at various nature camps in the Northeast who performed experiments with ants under his supervision. He describes on the first pages of his textbook how he buried blacksnakes in a box for six months of a Maine winter just to see what would happen to them (they survived).[1] Kinsey's passion for science as an undergraduate at Bowdoin College led him to enter graduate school in entomology, where his adviser suggested that he work on the genus *Cynips*. After he finished graduate school and became a professor at Indiana University (IU) in 1920, he taught biology education majors rudimentary human sex education beginning in the early 1930s. That teaching, along with changes in the landscape of taxonomy and biology, nurtured his interest in studying human sexuality more closely—and in cataloging and observing it every way he could.

The research that Kinsey conducted in his early career did indeed, as he later wrote to his graduate student advisee Ralph Voris (1902–40), determine his "entire future in research." His graduate education became the foundation for the investigative techniques and methodologies that he would refine over time for sex research. While Kinsey studied under William Morton Wheeler, one of the leading taxonomists of the Progressive Era, he learned to work hard in adverse conditions, to favor large sample sizes, to pay close attention to detail and variation, and to value observation and field naturalism over laboratory work. Wheeler also taught him that taxonomists could apply their skills to research on human problems, chiefly problems of sex and reproduction. Wheeler likewise modeled the persona of the cosmopolitan intellectual to Kinsey and to the other Bussey Institution students.

As Kinsey published the chapters of his dissertation and his first book-length essay, "The Gall Wasp Genus *Cynips*: A Study in the Origin of Species," in 1929, he affirmed his faith in taxonomy to reveal truths about the natural world, showed his bibliographic knowledge of taxonomy and of gall wasps, and began cultivating relationships with museum workers and other interested entomologists for the exchange of insect material, research questions, and ideas. At the same time, he developed an interest in biology education, principally sex education, in Indiana secondary schools. He became active in reforming courses and in teaching his curricular ideas to his biology pedagogy students. Kinsey's practical and theoretical learning experiences at the Bussey and his interest in sex education as a part of life science education provided the groundwork for Kinsey's move into human sexuality studies. Charting the history of Kinsey's gathering techniques, classificatory practices, and species descriptions shows how he created the epistemic categories and practices that became fundamental to his lifelong research methods.

Of Ants and Men: William Morton Wheeler and Taxonomic Practice

The life sciences were active fields in American academe when Kinsey began graduate study at Harvard University in September 1916. William Morton Wheeler and his cohort, many of whom received their PhDs from American universities in the 1890s, felt that their broad familiarity with the natural world allowed them to address human problems with authority.[2] Instructors at the Bussey Institution prepared Kinsey well for studying a variety of problems in the life and human sciences. Kinsey would be a member of the second generation of American biologists who felt qualified to speak out on human social problems and on their original areas of research.

Kinsey's early career in biology began inauspiciously. After two years of college at the Stevens Institute of Technology in Hoboken, New Jersey, where his father was an instructor, he transferred to Bowdoin College and graduated from there in 1916. He then enrolled at Harvard in the Bussey Institution's Graduate School for Applied Biology. The Bussey was first organized as an undergraduate school of horticulture and animal husbandry and began enrolling students in 1871. Wheeler, late of the University of Chicago, reorganized the struggling institution as a graduate school in 1908 with himself as dean. The Bussey was six miles from the main Harvard campus next to the Arnold Arboretum, and under Wheeler its focus shifted to training students in genetics and taxonomy. By 1929 it had granted sixty-seven master's degrees and sixty-three doctorate of science degrees, including Kinsey's. Of the 130 degrees that the Bussey conferred in its history, twenty-six students completed their degrees under Wheeler, mostly in the 1920s.[3] The Bussey—which remained unprofitable and at the literal and intellectual fringes of Harvard—closed permanently a year after Wheeler's death in 1936.

Kinsey was trained at the Bussey in conducting research under marginal living conditions and learned to think about insects professionally alongside colleagues with different objectives and methods. He also learned to value observation, mass collection, careful description, and precise organization of specimens. Kinsey long remembered that his graduate experience taught him to be adaptable. C. E. Hewitt, of the American Council on Education, wrote to Kinsey in 1931 asking him to supply a list of the skills and characteristics that taxonomic entomologists needed, so that Hewitt could publish them in the council's employment guide. Kinsey, then eleven years into his career as an IU zoology professor, thought that such a person "persists in his effort to solve insect taxonomic problems, usually in the face of disappointments and discouragement, and endures unpleasant conditions because of the purpose to be obtained."[4] He would indeed endure his fair share of "unpleasant conditions" while out in the field—and would push his entomological and sex research assistants to do the same.

Kinsey primarily studied with Wheeler and with Charles T. Brues, the two entomologists at the Bussey, and wrote his dissertation on the genus *Cynips*, or gall wasp. He also studied with Merritt Lyndon Fernald, a Harvard botany professor, with whom he would publish a book on edible wild plants.[5] However, as the whole school was in one three-story building, no students or faculty were far from one another. Kinsey's fellow graduate student Karl Sax, who later became a Harvard professor, remembered the toxic air in the badly ventilated animal laboratories: "I once placed a potted *Tradescantia* [spiderwort] plant in the rat room; the following morning it was black and completely dead." Most students lived in relative poverty in an underheated dormitory linked to the main building and killed, cooked, and ate lab rabbits when they were no longer needed for study, to save money on food.[6]

The students also had to navigate the different scientific philosophies of their

Figure 1.1. The Bussey Institution, greenhouse, and gardens in Jamaica Plain, Massachusetts, c. 1915. Donation by Edgar Anderson, Bussey Institution Records, Archives of the Arnold Arboretum, Harvard University.

professors. The genetics and taxonomy professors had divergent approaches to the treatment of data: for the former group, "the secrets of inheritance had come from artificially cultured forms," and the latter strove to categorize species in their natural states without human manipulation beyond a microscope, dissecting tools, and preservatives. Kinsey was thus engaged in scientific debates about the scientist's proper place vis-à-vis data early in his professional career, and he sided early on with the taxonomists rather than the geneticists. The instruction he received, most notably from Wheeler, emphasized collection and description over all laboratory methods except for the most nonintrusive forms of experimental gall wasp breeding.[7] According to Sax (and confirmed by another Bussey student, Edgar Anderson), genetics students like himself did not interact with Dean Wheeler much, as he "was busy classifying and studying ants most of the time." In Wheeler's early career, he had devoted his attention primarily to ant taxonomy and to the description of new species. While Wheeler never left taxonomy entirely behind, he eventually became more interested in insect behavior, convinced that behavior influenced insect physiology and anatomy. His attempts to craft animal behavior into a science separate from biology did not succeed.[8] Wheeler also had a strong interest in human sexual and reproductive behavior, an interest that his most famous graduate student would later share.[9]

As a graduate student at the Bussey, Kinsey boarded in an elderly widow's home nearby, and, according to his botanist friend Edgar Anderson (1897–1969), he kept himself somewhat apart from the other students, saying that Kinsey "consciously acted as though he were a little better than the others and most of them didn't like it. He never had any fusses or anything of that sort, just smiled at them in the hallway. I thought he was better than any of the rest of us."[10] Though there was not the same level of material sharing among the taxonomists as there was among the geneticists, Kinsey nonetheless began to cultivate the relationships he would need to maintain his own gall wasp collections. The poor conditions at the Bussey bonded students. Sax, a murine geneticist, remembered a catchphrase that circulated around the badly equipped Bussey Institution: "If a student could do research at the Bussey he could do it anywhere." Anderson concurred, stating in a 1962 interview that the institution's lack of well-equipped facilities inspired students to be creative in their work. "You learn by what is essentially the apprentice method with a minimum of courses and a minimum of red tape and minimum of equipment. . . . Anything that was badly needed you got it; otherwise you didn't." Kinsey, like Anderson and Sax, learned from his experience at the Bussey and from studying under Wheeler to work with minimal apparatus and to depend on his own skills to plan and to develop his research with little assistance from others.[11] Anderson well remembered Wheeler's hands-off approach with his graduate students: "I was working in Dr. East's laboratory one morning when Dean Wheeler . . . came in with Kinsey's thesis in his hands. 'It's a remarkably fine piece of work' he said to Dr. East, 'he really is of high caliber.' All of the Bussey professors left the students pretty much to their own devices but Wheeler as a matter of principle carried this policy to greater lengths than any of the others. He really had seen so little of Kinsey that he had not sized him up until the graduate years were practically over."[12] Though Wheeler was remote from Kinsey's everyday work, Kinsey absorbed through his adviser's lectures many aspects of his views on the role of scientists in managing their own careers and in studying human behavior.

When Wheeler held leadership positions in biological organizations, he gave two significant addresses that provided his perspective on scientific practice: "The Dry-Rot of Our Academic Biology" and "The Termitodoxa, or Biology and Society," both given at American Society of Naturalists (ASN) meetings. Wheeler presented "The Termitodoxa" at an ASN symposium in December 1919, illustrating that "finding ruthless analogy between ant and human appealed to Wheeler's acerbic wit and was one of his chief pleasures in myrmecology."[13] He read an imaginary letter from King Wee-Wee of the 8,429th Dynasty of the Bellicose Termites to the human race. Wheeler moved textually back and forth between the insect kingdom and the human world, arguing through the voice of the termite king that the insect and human worlds were compatible, if not interchangeable. The lecture was undoubtedly funny to his listeners and to his readers, but there was plenty of real social criticism embedded in it that Kinsey registered.

In the "Termitodoxa," Wheeler argues that society is an artificial by-product of "termite" interaction, and a multiplicity of forces had corrupted it in the past. In the voice of the termite king, he gives a scathing critique of the negative impact of priests, pedagogues, politicians, and journalists on "termite society," not to mention "the commonwealth anarchists, syndicalists, I. W. W., and bolsheviki . . . profiteers, grafters, shysters, drug-fiends, and criminals of all sizes interspersed with beautifully graduated series of wowsers [social reformers], morons, feebleminded, idiots, and insane." Society was in trouble until another termite king "had the happy thought to refer the problems of social reform to the biologists." Biologists reconfigured termite society into a "superorganism," so that every member of society had a prescribed role. Population and food supply were regulated, with reproduction restricted to the highest and healthiest caste. "Rigid eugenics at the same time," Wheeler proclaims, "solved the problems of ethics and hygiene, for we were thus enabled, so to speak, to ram virtue and health back in the germ-plasm where they belong." Biologists, for Wheeler, were best suited to design social policy in the human realm, and human society would be healthier and more smoothly run if they were in charge. While Kinsey would not advocate eugenic thought in either *Sexual Behavior in the Human Male* or *Sexual Behavior in the Human Female*, he clearly took to heart the idea—indeed, the responsibility—to study and to speak out on human social problems.[14] He likewise agreed with Wheeler about the negative influences of religious and social reformers who sought to change peoples' behaviors to meet their own artificially constructed standards.

Wheeler gave "The Dry-Rot of Our Academic Biology," another address that resonated with Kinsey, as a pep talk of sorts as president of the ASN at its December 1922 annual meeting. He intended his address to encourage professional biologists to share the love of their field with amateurs, college students, and high school students in order to cultivate their nonprofessional lifelong interest in the natural world. He told his audience that real biologists were naturalists, not laboratory geneticists. As Charlotte Sleigh points out, in Wheeler's thinking, "laboratory-based scientists never saw nature red in tooth [and] claw; neither did they suffer the deprivations of fieldwork that might cause them to recognize related drives in themselves, drives whose management cultivated manly character."[15] For Wheeler, field naturalists were strong and masculine, and laboratory researchers were effeminate and overly interested in their microscopes. Wheeler thought that basic behavior and morphology were the best subjects for (presumably male) graduate and undergraduate college students "without cytological lace and ruffles." Biology teachers and students should spend a good amount of time in the field, especially abroad, he argued, as it kept their work and perspective fresh. Nurturing the enthusiasm normally found in amateurs and in their young students was the key to better biology research practice and to better teaching: "We should realize, like the amateur, that the organic world is also an inexhaustible source of spiritual and aesthetic delight."[16] According to Sleigh, he sought a form of practice "that respected the methodological freedom of the naturalist, unaffected by the fads of academe and the laboratory."[17] Wheeler received two dozen letters after presenting his address, including letters from the statistician Raymond Pearl and the psychologist Robert M. Yerkes, all praising Wheeler for his wit, candor, and facility with language.[18]

Wheeler's esteem among his peers did not extend to all of them, chiefly geneticists. Wheeler and the entomologist Richard Goldschmidt, later a foe of Kinsey, did not get along, especially after Goldschmidt had an unpleasant visit to Wheeler's laboratory. The Bussey professor Edward M. East confirmed Goldschmidt's feelings about Wheeler in a 1924 letter: "I had a quarrel with him in 1917 and we did not speak for over a year. He has made it very unpleasant for me. He hates all experimental work." East wrote Goldschmidt again two months later that Wheeler's poor treatment of Goldschmidt "was probably not personal but part of his over compensation against the progress of genetics. He hates genetics simply because it has gotten further and faster than the stuff he is interested in."[19] It is impossible to know if Kinsey knew about Wheeler's treatment of Goldschmidt during the latter's visit, and if that treatment affected Kinsey's and Goldschmidt's professional rivalry, but he may have sensed that Wheeler stalled his own academic career in the 1920s by not embracing genetics as an important new aspect of entomological study. In other words, "Wheeler's problem was a sometimes willful confusion of field science and 'mere' natural history on the part of his laboratory-based colleagues, the emerging elite in biology."[20]

Kinsey would adopt into his own worldview some elements of Wheeler's thought. He certainly agreed with Wheeler's idea that engaging students with the natural world was an excellent means of forming them into lifelong amateur naturalists and biologically aware citizens. His teaching guide, textbooks, and workbooks all emphasize the need for instructor engagement and hands-on learning. Kinsey quotes Wheeler's title phrase in his pedagogy guide, *Methods in Biology*: "The teacher who would avoid the dry rot of teaching must expose himself, as well as his students, to a variety of methods."[21] Also, Wheeler's gendered separation of field and laboratory work into real/masculine and lightweight/feminine forms of science, respectively, clearly struck a chord with Kinsey. Wheeler's characterization of fieldwork as masculine and his own persona may have appealed to a young protégé hoping to establish himself as a "real" academic biologist.[22] Wheeler showed Kinsey "how an ethos must be grafted onto a scientific persona, [how] an ethical and epistemological code [could be] imagined as a self."[23] But Kinsey would not follow Wheeler's example regarding his treatment of his colleagues. Kinsey tried to maintain businesslike if not cordial relationships even with his enemies. However much Wheeler influenced Kinsey's investigative strategies, relationship strategies, and any accompanying gendered characterizations, Kinsey's disinterest in pursuing research with standardized animals in laboratories rather than with those gathered in the field would ultimately be part of his decision to cease researching gall wasps and to commence his human sexuality research.

As for Wheeler's ideas in "The Termitodoxa," Kinsey too would not avoid comparing the complexity of social insects' worlds to human society. He would harshly criticize contemporary American political, medical, and religious leadership on social problems in the *Male* and *Female* volumes. He had faith that a scientific, objective, nonjudgmental approach could articulate and solve sex-related social problems better than other mechanisms could. While he entertained eugenic ideas in an address to Phi Beta Kappa students at IU in 1939, he would never embrace the radical program of eugenics that Wheeler—however playfully—advocated.[24] Instead of promoting the elimination of undesirable persons from society, Kinsey would strive to include as many as possible in his sex history interview data to diversify his sample. While Wheeler used mass collecting to develop theories about categorizing those he considered undesirable or unproductive so that he and other scientifically minded individuals could direct their improvement, Kinsey used mass collecting to show how harmful such categorizing was for those engaged in nonnormative sexual practices. Kinsey refrained as well from suggesting that they should change their sexual behavior.

Despite Kinsey's disagreement with many of Wheeler's ideas, Kinsey enjoyed sharing Wheeler's speeches with others. He references "The Termitodoxa" in the outline for his general biology course in 1928. Louise Ritterskamp Rosenzweig, a student worker in Kinsey's gall wasp lab, remembered that it was one of his favorite pieces of writing. "He would chuckle over it at his desk and periodically lend a reprint of it to some graduate student with the admonition that he treasured it." He would tell the borrower, "Please do not forget to return it."[25]

Though the purposes and claims of Kinsey's and Wheeler's practices eventually diverged, Kinsey's early training as a classical taxonomist under his adviser was standard for the time. For Wheeler, "classification was pointless and dull when considered in isolation from animals' conduct, and conduct was meaningless without a grasp of phylogeny."[26] As "the objective of traditional taxonomic investigation is the orderly and systematic organization of knowledge about the biological world," both Wheeler and Brues encouraged their students to focus on morphology, to learn how to study each individual insect and its parts closely, and to record their information separately.[27] Bussey teachers and students made their way through numerous insect genera, including Hymenoptera—the order on which Kinsey eventually decided to focus. Wheeler and Brues trained him to separate insects into parts—either visibly or with instruments, and often using a microscope—for identification and classification. As it can be hard for a scholar to know "which differences are salient for differentiating entities from one another and which ones are only negligible differences among variants of a single entity," Kinsey and his fellow students needed and received much practice in figuring out which differences mattered, and which did not, in large insect sorting operations.[28]

Thus one of Kinsey's earliest and lasting knowledge-making practices was to separate the evidence of a single entity into distinct parts and to record each of the data points individually for later examination and reassembly next to like data. He then transformed his findings into orderly text. That textual description could serve as an exercise in the identification of species or could lead to the establishment of new ones. As "by the grouping and regrouping of his data . . . the scholar discovers new relationships, new approaches to old problems, and new areas for exploration," Kinsey began learning in graduate school how to group and to separate data points in order to clarify their relationships and perhaps to discern cause and effect among them. Gall wasps and their component parts, after collection and organization, became epistemic objects. The right grouping of data from them might just lead to new discoveries in evolutionary science along with new species. "The trick of good sorting," Robert E. Kohler has noted, "is knowing which variations are significant taxonomically and which ones are just random noise."[29] Kinsey needed to learn how to make visible the gaps between species—but first he had to gather enough gall wasps to make those distinctions properly, and to learn the difference between taxonomic significance and "random noise."

Wheeler and Brues taught Kinsey to be a thorough and efficient taxonomist and imparted to him their ideas about the relationships between science and society inside and outside the classroom. Kinsey took four graduate courses during his three years at the Bussey: General Zoology 7a and General Zoology 7b in the 1916–17 school year, and General Zoology 20f, taught by Wheeler, and General Zoology 7c, taught by Brues, from April to June 1919. In the General Zoology 7a and 7b courses, Kinsey was one of only three students, and in General Zoology 20f, he was one of five students, including the first female Bussey student, Esther Hall.[30] He received A grades in the first two zoology courses, but his grades for the latter two courses are not recorded. The former Bussey graduate student Leslie C. Dunn remembered that the graduate courses met once a week, and each week the professors assigned students readings that they would have to summarize and to defend. Usually Wheeler and East challenged the students to think analytically about their arguments, and Wheeler tended to be the most severe critic of students' work. Another former graduate student, Frank Carpenter, thought that "Wheeler was the world's worst lecturer" who digressed from his notes too often.[31] Regardless of what Kinsey thought about Wheeler as a lecturer, he took copious notes on the history, principles, and application of taxonomy in Wheeler's courses.

The ideas in Wheeler's lecture on "Taxonomy of Insects" in Zoology 7a would have long-term reverberations in Kinsey's scholarship. Taxonomy was particularly important for entomologists, Wheeler said, "otherwise entomologists would never be able to see the forest on account of the trees." Wheeler emphasized that classification of the natural world had a long history with many imperfections, and students needed to base their classifications on the whole animal, not just on one or two parts: *"All the characters must be weighed* and considered before placing an insect in a particular group. Other things equal, the most recent and sanest classification in this respect is the best, but no classification is final. This is clear from the fact that we do not know a fourth of the species of insects now living on the planet." Furthermore, taxonomists needed to be aware that classification standards changed over time, and that future scholars might modify their species designations. In the end, individuals "are really all that are immediately given in nature. All other categories are subjective and become more and more indefinite and saturated with the opinions and estimates of individual entomologists as we mount upward."[32]

What Kinsey learned about entomology, and his developing thoughts about science in general, are evident in his surviving zoology course notes. He interspersed his notes for Zoology 7a and Zoology 7c with copies of the intricate drawings of insect embryos, nervous systems, sense organs, and brains that Wheeler drew on the chalkboard. He occasionally made offhand comments to himself as well. On the first day of Zoology 7a, Kinsey noted, "Insects & man are only contestants for possession of earth. Will man or insects conquer? . . . Small size of objects of study—leads to narrowmindednest [sic] in Entomologists. He should have a bigger side interest."[33] Kinsey was a devoted entomological student, but from his time in graduate school onward, he knew that he wanted to cultivate other academic interests as well. He also realized that much of the world's flora and fauna awaited study and classification by trained zoologists like himself. He began thinking through the broader implications of having the ability to lump and to split individual insects into groups and to name species using taxonomic practice. He wrote late in the course in his notebook, "Plenty of work left to do. Without attention to classification Entomologist would be lost in minor details." The ability to regard items "as typical members of categories is a prerequisite for classifying any group of phenomena," and Kinsey certainly needed that skill for understanding the often minute morphological differences between gall wasp species. He left the class convinced of Wheeler's point about the importance of the individual specimen to taxonomic study, which he would repeatedly return to throughout his research career: "Individual is by far the most important thing—the only thing given. Some people consider all groups above individual as purely imaginary—but not strictly so."[34]

Mentioning the importance of the individual specimen and stating his ambivalence about higher categories foreshadowed his later disavowal of higher categories in his last book on gall wasps, *The Origin of Higher Categories in Cynips*. That brief comment also hints at his realization that learning to identify, describe, and categorize insects produces not just biological facts, "but the very categories by means of which we perceive and understand our world."[35] By learning taxonomic technique through insect description and classification, Kinsey began to explore the classifications by which he ordered his own perceptions of the world. As learning to classify broad swaths of natural phenomena meant learning how human beings gave meaning to the natural world as they experienced it, Kinsey therefore embarked on his own lifelong research project to articulate the world's divisions as he distinguished them. The humble cynipid would be his first focal point for doing so. While gall wasps were an excellent choice of genera for a scientist as attuned to small details as Kinsey, they would become problematic as study species when the questions and methods of academic biology began to change.

The Way to Collect

Kinsey wrote his doctoral dissertation, "Studies of Gall Wasps (Hymenoptera, Cynipidae)," on a tiny insect—most about two millimeters long—that, in his opinion, entomologists had understudied. Female gall wasps lay eggs on host plants, usually oak trees or rose bushes. A chemical reaction causes the plant to form a gall, which grows to various shapes and sizes depending on the species of wasp and tree or plant. The larvae mature inside the gall and eat its inside layers as food until they are strong enough to chew their way completely out. The galls themselves are made of three to five layers of plant tissue, again depending on the species of wasp and host. After emerging, the adult gall wasps live one to three days, just long enough for the females to reproduce.[36] As gall wasps had not been a major focus of other entomologists' study, Kinsey was able to pioneer their scientific classification using the intellectual tools of the twentieth-century taxonomist.

After Kinsey defended his doctoral dissertation on September 25, 1919, he spent a year traveling alone throughout North America on a Sheldon Traveling Fellowship that Wheeler helped him obtain.[37] His aim was to survey the continent to further develop his own collection of galls and to gain an idea of gall wasps' geographical variation. His earliest collecting trip was undertaken on foot with occasional rides on trains and buses. He regularly sent his findings back to the Bussey Institution, where his former graduate colleagues would mind them until he returned. When the insects (and their parasites) emerged from the galls, he had to trap and to kill them, then organize and store them. The wasps hatched from inside the galls in the Bussey laboratory at least once before Kinsey reappeared. Edgar Anderson told Kinsey that he "vividly recalled the stew that Brues got into when the Cynipids [Kinsey] sent from the west coast emerged and filled the buildings!"[38]

Kinsey collected approximately three hundred thousand specimens during that yearlong sojourn. They took him months to classify and provided him with a wealth of material. Those numerous specimens enabled him to nurture both his intent to focus on isolation as a mechanism of speciation and his desire to make his collections even more complete. As early twentieth-century taxonomy depended "as much on geographical and ecological knowledge of living creatures as on physical measurements made on preserved specimens," Kinsey was delighted to get out into the field and to make his own contributions to science based on his firsthand observations.[39] In March 1920 he wrote to Fernald with excitement from Merced, California, about his newfound ability to identify properly the oak trees in which gall wasps lived: "When I first met with these evergreen oaks, I found it quite bewildering to attempt to distinguish species by means of leaves. But I soon found that the species were all very distinct, and when one *in the field* has the whole tree, its various parts, its habit, etc., to go on, it is easy enough to feel certain of the species."[40] Collecting in the field concretized Wheeler's classroom lesson about taxonomists needing to see whole biological entities in order to understand and to identify their parts.

In correspondence throughout the yearlong trip, Kinsey was notably enthusiastic about being the leading entomologist to classify North American gall wasps. He mused in the same letter to Fernald, "I am fortunate in being the first collector to work on this group in many of these states; and the number of entirely distinct faunas here in the southwest, makes the original exploring almost as rich as tho [sic] one were entering a previously unknown continent."[41] As "sorting specimens by geographical location was often as revealing as sorting by physical differences, especially where the latter were small," as they were in gall wasps, if Kinsey were to make a valid and lasting contribution to entomology, he not only wanted to be the first entomologist to classify in the genus *Cynips*. He wanted to do most of the collecting and sorting himself.[42]

Kinsey described his 1919–20 collecting trip to Natalie Roeth, his South Orange, New Jersey, high school teacher who had encouraged Kinsey to pursue his interest in science as a teenager. His letter to her captured an ideal of survey collecting that matched the enthusiasm of his peers, who also gathered and sorted numerous specimens in zoology and botany:

> I spent the time wandering over the country, collecting insects, especially the gall wasps on which I have been working. . . . Think of that for a life! I am more and more satisfied that no other occupation in the world could give me the pleasure that this job of bug hunting is giving. . . . [I traveled] as a biologist, wearing khaki, living with my all on my back in my pack, getting off into the wildest parts of the country, often into regions where few people ever get. Inasmuch as oaks, on which the gall wasps occur especially, are found only in the highest mountains of most of the country, I got into practically every one of the mountain ranges except those of the far north. In one range in Arizona, for instance, I got off where for four whole days I didn't see a solitary man; I was about fifty miles from the nearest town, living on my own camping wit. . . . In all I covered about 18,000 miles, of which 2,500 were on foot.[43]

In an age when entomological, botanical, and many other specimen collections were still too sparse to make accurate geographical or morphological delineations of species, Kinsey was far from alone in his desire to gather for himself the most thorough and comprehensive body of data that his time and energy allowed. Many other life scientists racked up thousands of miles traveling and gathered wide collections of specimens as a means of understanding some part of the world. As Peter Galison and Lorraine Daston wrote regarding scientists of this era, "the will-based scientific self was articulated—built up, reinforced—through concrete acts, repeated thousands of times."[44] As he continued his extensive collecting and sorting, Kinsey continued to fashion himself as a scientist through his focused work ethic and academic persona. Kinsey's dedication to completing the repetitive nature of taxonomists' tasks fashioned his scientific self.

Kinsey's gathering trips around the United States in the 1920s, and then to Mexico and Guatemala in 1935 and 1936, broadened his perspective and ambition. His trips mostly took place in the late summer and fall, as at that point the galls containing the wasp larvae were mature enough to survive the harvest from the trees or plants on which they grew and the journey to the laboratory where the breeding took place. Kinsey considered sampling in a single location "complete only when the analysis of new individuals failed to change the average measurements recorded for those already on hand."[45] He often took research assistants along on those trips, who well remembered how Kinsey organized his collecting days. His graduate student Robert Bugbee, who accompanied Kinsey on several trips, recalled that after a full day's collecting, Kinsey and his assistants would spend "hours after dinner . . . sorting and labeling the collections of the day, [which] assured maximum utilization of time."[46] Kinsey's great enthusiasm for their discoveries and for teaching students how to collect somewhat

Figure 1.2. Alfred Kinsey inspecting insect galls, c. 1930. Courtesy of the Kinsey Institute for Research in Sex, Gender, and Reproduction, Inc.

mitigated the exhaustion each one felt at the end of an expedition. He was also attempting to teach them how to be scientists after his own fashion. If "the mastery of scientific practices is inevitably linked to self-mastery, the assiduous cultivation of a certain kind of self," Kinsey desired to teach his students taxonomic skills and also to model his ideal of the proper way to have a scientific self.[47]

Following the mass collecting trips with his graduate students, Kinsey needed to preserve and to organize the specimens for long-term storage. He stored his insect collection in Schmitt boxes, wooden boxes with foam interiors and glass lids that slid off for easy viewing. He built the insect collection over twenty years, and while he did have student help with its maintenance, he did much of the work himself. Anderson, who visited Kinsey's laboratory in Bloomington several times throughout the 1930s and 1940s, remembered the details of his laboratory organization process: "Each insect was glued to a snippet of stiff paper and impaled on a steel pin. Minute paper labels . . . were affixed to the same pin and thousands upon thousands of these exquisitely prepared specimens were pinned in . . . boxes. The specimens were examined (and frequently re-examined) under a dissecting microscope. If they belonged to a known species and variety, they were so labeled; if not, the proper pigeonhole was worked out."[48] The careful organization of the gall wasps was predicated on their being labeled and stored very precisely. All of Kinsey's employees, mostly female undergraduate students, followed specific procedures for mounting the gall wasp specimens. Kinsey's former student June Hiatt Keisler remembered them clearly decades later:

> To mount a gall wasp, one picked it up carefully, with tweezers, applied a transparent cement to the left side of the insect, and placed it on one of the clips of cardboard, which was then impaled on a two-inch-long steel pin. The cardboards were mounted two deep and about three-quarters of an inch apart on each pin, and the pins were set about a quarter of an inch apart in cork-bottomed boxes and in rows about two inches apart. The important thing was that the insects must not be broken and that they must all face toward the right. . . . I learned to watch the window-screen collecting envelopes in the boxes outside the office windows for the emerging adult insects, to prepare and use a killing bottle, and later, using Dr. Kinsey's special lettering technique, with India ink to write minute, legible collecting data to label each insect.[49]

Another employee recalled that the labeling required a magnifying glass on which Kinsey taught assistants like herself to "print perfectly . . . on this *teeny tiny*" label. That level of precision was necessary to maintain a large collection of insects and galls for long-term storage and easy reference. His assistants were also trained to watch for the gall wasps that had not yet emerged from their galls that were incubating on tree limbs or rose bushes inside the laboratory. Given the gall wasps' short life span, it was especially important that Kinsey not miss the opportunity to observe their hatching and mating. As Joe Cain has pointed out, "Kinsey enthusiastically used 'experimental' techniques to help with his routine nomenclature decisions."[50]

However, Kinsey's research operation not only included the stabilization and organization of the wasp collection; it also required an entire laboratory system of classification that necessitated the management of complex interactions between material collections and texts. As Anderson put it, "hundreds of [Schmitt] boxes, together with field notebooks, collections of the actual galls, [and] herbarium specimens of the oak trees from which the original collections were made were kept in orderly storage for effective cross reference."[51] Each of the boxes held approximately eight hundred gall wasps, for each of which Kinsey and his assistants took twenty-eight separate measurements, which in turn led to yet another organizational and recording mechanism that shaped how Kinsey manipulated and analyzed data.

Wardell B. Pomeroy joined Kinsey's research team in 1941 as Kinsey's work on the gall wasps was waning, but he remembered how the older man recorded the data for each gall wasp specimen. "It was laborious work," he recalled. "For this purpose he began to develop a shorthand system he had invented, using small symbols. Out of this experience came a coding device. . . . Translated to the uses of sex research, it enabled us to gather the equivalent of as many as twenty to twenty-five typewritten pages of interview data on a single sheet." None of the entomology data recording sheets survive, and one of Kinsey's more recent biographers disputes the importance of this recording sheet.[52] However, Pomeroy's remembrance of them provides a critical link between Kinsey's entomological and sexological research, and indeed joins together Kinsey's methods of organization for insect and human data.

Pomeroy's memory of the entomology recording sheet shows that Kinsey had agreed with Anderson's critical discovery in his extensive study of irises: that "what we need for the species problem is a method whose fundamental observations are based upon the qualitative categories of taxonomy but which treats these categories in such a way that they can be used for comparison and analysis."[53] The gall wasp recording sheet facilitated the recording of qualitative data about individual insects for quantitative analysis. It was likewise a tool for gathering all of an individual gall wasp's data on a single sheet using easily learned and accessible shorthand. The gall wasp recording sheet also led in part to the development of Kinsey's interview sheet for the sex histories—and the practice of their recording—that he would begin to take in the summer of 1938 during the first session of the IU marriage course. Clear and orderly management of multiple data collections in various media and the integration of those data collections for the creation of new scientific data were central to Kinsey's research practices and knowledge production throughout his academic career. As Edgar Anderson recalled:

> The urge to build up a significant collection is a special sort of inner drive. . . . In Kinsey the strength of this compelling inner fire showed itself increasingly and in various ways, when he closed one of the tight-fitting insect boxes and put it back on the shelf in the proper place, when he inserted one of his coded sex-survey data cards and closed the steel filing case, there was a physical reaction which I have noticed in other scholars and collectors. The box lid was not merely closed, it was slowly but deftly pushed shut and the tension of the fingers showed that the closing was of some inner significance. When the drawer of the filing cabinet was pushed shut the fingers lingered on the drawer until it slid firmly into the closed position and there were meaningful tensions of the arm and back muscles. Without [the urge to collect] we should never achieve these vast stores of codified information which are one of the prerequisites for a scientific understanding of the world.[54]

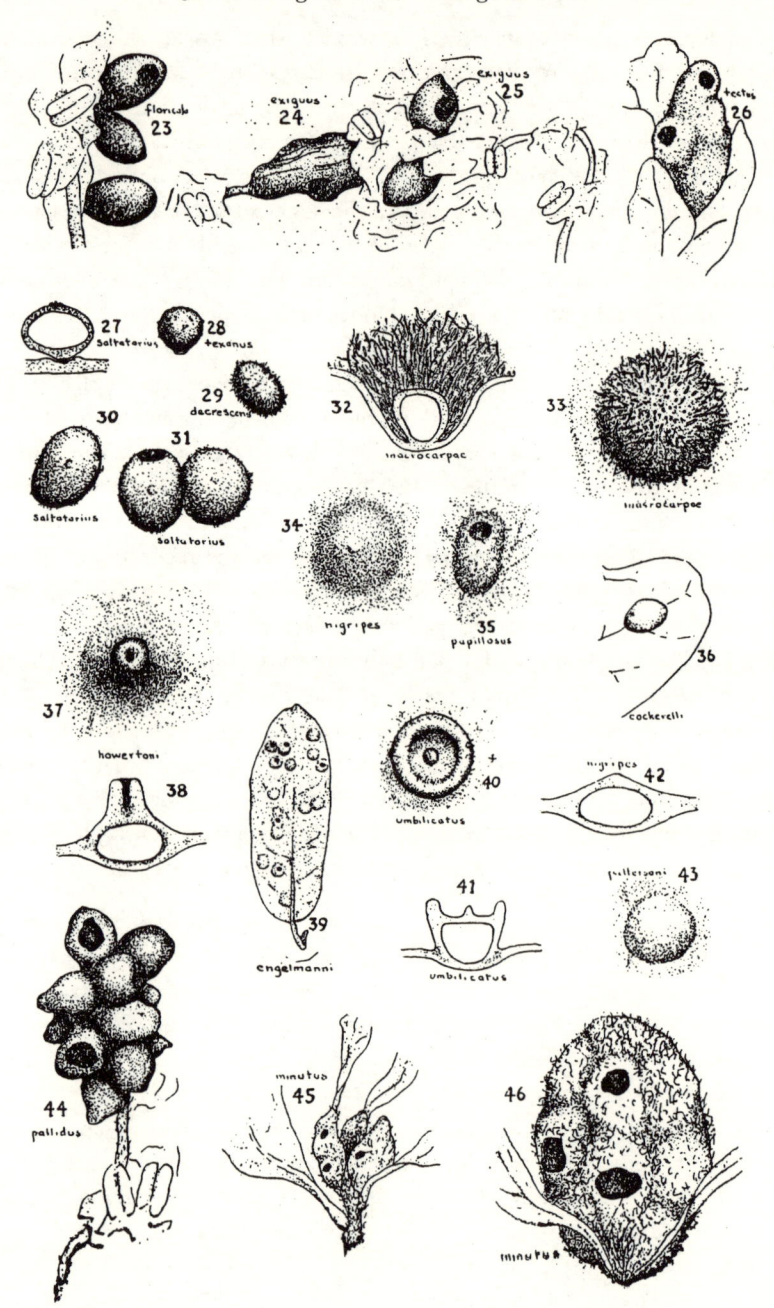

Figure 1.3. Alfred Kinsey's drawings of galls formed on various types of trees and plants. Alfred C. Kinsey, "The Gall Wasp Genus *Neuroterus* (Hymenoptera)." *Indiana University Studies* 10, no. 58 (June 1923): 145.

Anderson vividly remembered Kinsey's physical and emotional attachment to his collections and storage mechanisms. He located Kinsey among other collectors who made evident the deep personal meaning of their collections through their interactions with the objects in those collections. The collection had a notable smell of mothballs and other preservatives used to maintain the collection whenever the lid was off, so Kinsey and his associates had to interact with the insect collections on multiple sensory levels. Anderson's observation of Kinsey demonstrates Galison and Daston's point that "scientific objectivity resolves into the gestures, techniques, habits, and temperament ingrained by training and daily repetition. . . . It is by performing certain actions over and over again—not only bodily manipulations but also spiritual exercises—that objectivity comes into being."[55] Kinsey showed in his deep connection to his insects and their storage boxes his passion for both the objects themselves and the scientific practices that ordered them. He would bring that drive for best objective practices to building and to maintaining his relationships in the entomological community as well.

After Kinsey attained a professorship at IU and moved to Bloomington in 1920, he continued to take short trips, to solicit galls and gall wasps from other collectors' and to publish articles on his new findings throughout the late 1920s and early 1930s.[56] He began to network with other entomologists around the world as well, far beyond the museums that had lent him specimens for his dissertation. He traded, gave, or sold gall wasps to others, encouraged fellow entomologists to visit him, and organized statewide meetings of Indiana entomologists.[57] Some sent him gall wasps to identify. Kinsey established himself early in his career as a reliable taxonomist engaged in the rounds of exchange that characterized entomological professionals at the time.[58]

Kinsey consciously fostered the professional relationships that would help him in the future, primarily by giving away parts of his collections to institutions that had permitted him to borrow theirs and whose publication divisions printed his taxonomies. His proffered thanks to the curators of those three museums is one indication of his entry into the give-and-take world of academic science. Publishing Kinsey's chapters in the *Bulletin of the American Museum of Natural History* was perhaps a favor that the museum staff granted to him because of Wheeler's close relationship to—and patronage of—the museum. Kinsey also offered to sell the gall wasps that he studied and owned to the American Museum of Natural History (AMNH), but he later decided to donate them instead, presumably as a gesture of goodwill toward the *Bulletin* staff. Kinsey promised as well to give material to the AMNH in the future, and he did so in 1930 and 1931 after the publication of "Gall Wasp Genus *Cynips*."[59]

The AMNH entomology curator did not think much of Kinsey's early academic efforts. Of Kinsey's first donation, he sniffed, "This gift is a result of a

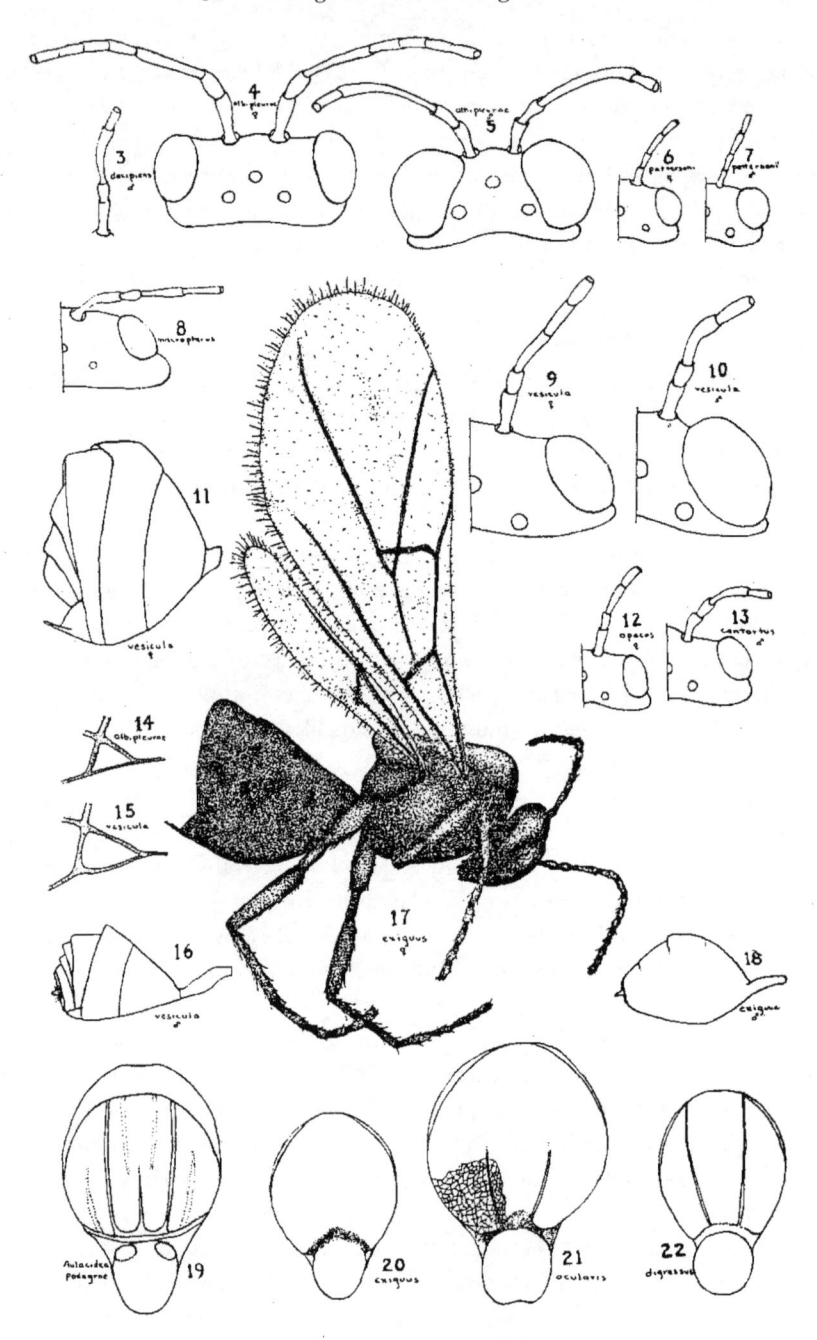

Figure 1.4. Alfred Kinsey's drawings of gall wasp antennae, heads, abdomens, areolets, and thoraxes. Kinsey, "Gall Wasp Genus Neuroterus," 143.

poorly founded hope that we will publish for him again sometime." Nathan Banks, the entomology curator at the Museum of Comparative Zoology (MCZ) at Harvard, expressed his disappointment that Kinsey did not publish his dissertation chapters in the MCZ's bulletin, and he promised to promote Kinsey's work to its editor the next time Kinsey had material to publish. Kinsey chose not to publish his work in the MCZ bulletin, but he did donate some of his specimens to the museum in 1922. As Kinsey began publishing papers in institutional publications and making donations of specimens, he became active in those quotidian processes of academic life and aware of their politics. He also ensured that both his specimens and the taxonomies he derived from them would become a recognized part of the broad entomological survey collecting in early twentieth-century North America.[60]

Kinsey's wife Clara donated his gall wasp collection to the American Museum of Natural History in 1957, a year after his death, and Paul H. Gebhard donated the accompanying notes that he found in the Institute's office files three years later. Gebhard sent "a quantity of notes and other materials dealing with gall wasps" in two cartons, where they presently remain. The estimated size of the Kinsey gall wasp collection at the AMNH ranges from 5.5 million to 8 million individual specimens. It has never been completely counted or cataloged.[61] Kinsey's massive collection, established by his early museum use and donation practices, would cement his long-term place in entomology.

Becoming an Entomologist

Kinsey began his published contribution to the life sciences, as did many of his peers, with his doctoral dissertation. He describes sixteen new gall wasp species in its three chapters, which were later published nearly verbatim in the *Bulletin of the American Museum of Natural History*. He also outlines the life histories of ten previously studied species and theorizes the evolutionary relationships of three cynipid genera. Kinsey used specimens from the AMNH, the Boston Society of Natural History, and the MCZ at Harvard, and in the acknowledgments he thanks the curators of each for unfettered access to those collections.[62]

The first chapter of the dissertation, "New Species and Synonymy of American Cynipidae," contains lengthy descriptions of the sixteen new gall wasp species. This example is Kinsey's description of the female *Neuroterus thompsoni*:

> FEMALE. Almost entirely black, the antennae ringed with light yellow at the second to third joint; length under 2.0 mm. HEAD: black, mouth-parts reddish brown; antennae 13-jointed (inclined to curl in dried specimens), joints one and two stouter than the following joints though not as globose [spherical] as in most species of *Neuroterus*, first joint dark brown, part of the second joint and the proximal tip of the third joint very light yellow, joints

three to thirteen dusky brown-black. THORAX: piceous [pitch-colored] black; mesonotum [back] microscopically cracked, without grooves; sides of the thorax lighter reddish black. ABDOMEN: piceous black, subpedicellate, rather angulate [angled] in outline; ovipositor apparently short. LEGS: light brownish yellow, the hind femora and tibiae dusky brown except at the joints; tarsal claws almost black. WINGS: veins brown, the cubitus [major vein] reaching the basal vein below the midpoint, the areolet rather large, the radial cell long, narrow, open. LENGTH: 1.7 mm.

Such a detailed definition not only served to describe the species; it also allowed Kinsey to reorder the genera properly and to show each species' genetic relationship to the others.[63] His correlation of large sample sizes with increased accuracy in species identification and description is clearly evident in chapter two, "Life Histories of American Cynipidae," and throughout the dissertation. In that chapter, Kinsey continuously shows his preference for observation in the field over the lab and his attention to detail, by adding to extensive descriptions of species a bibliography of any previous scholarship on the species, including notations of previous scholars' errors, and a discussion of the life cycles of the insects.[64]

Kinsey's dissertation work was based on his own collection of cynipids as well as on those of existing American museum collections and some specimens borrowed from Europe. He gives lengthy descriptions of male and female type specimens broken down into their important components and then refashions them in textual form for scholarly use, as any well-trained taxonomist would do. He corrects some previous scholars and describes new species. Kinsey's dissertation and first publications show that he had become one of the classical taxonomists that Wheeler and Brues expected their students to become. They also show how alike Kinsey's collecting processes and taxonomies were to those of his colleagues. Kinsey was inclined toward "splitting" in his species creation—drawing species narrowly with few variations. In an early work, he added an additional layer of division between species and genus to give his taxonomy the specificity he wanted: "By the employment of varieties I am open to both the charges of being a 'lumper' and a 'splitter,' the former in my treatment of species, the latter in my recognition of varieties; but in any event the scheme is employed with consistency, and does portray the different degrees of relationships that actually exist."[65] Kinsey's 1920s-era publications, with their lengthy lists of minute descriptions of new species, show Kinsey's tendencies toward splitting species at that stage of his gall wasp collecting. As Robert Kohler has noted, "periods in which collecting is expanding but not yet comprehensive will favor splitters, because incomplete knowledge of natural variability makes it easy to see a few distinctive individuals as representatives of true species."[66] "Lumping" species could be done after survey collection for a genus was complete, which was not yet the case for gall wasps, nor for many other North American flora and fauna.

Kinsey published several articles based on short collecting trips throughout the 1920s, and his first full-length entomological text, "The Gall Wasp Genus *Cynips*: A Study in the Origin of Species," appeared in 1929.[67] "Gall Wasp Genus *Cynips*" shows that he was not alone in his preference for using large numbers of specimens for increasing accuracy. His numbers for that text are large (93 species, 17,000 individual insects, and 54,000 galls) but not excessive, especially in comparison with other taxonomists working on different types of species. He wrote of his fellow taxonomists, "we believe it no coincidence that our conclusions more nearly accord with those of [Emmett Reid] Dunn who studied 12,600 specimens of the 86 forms of the salamanders of the family Plethodontidae, or of [Clarence Eugene] Mickel who studied approximately 10,000 specimens of the genus *Dasymutilla* [velvet ants]." His colleague and friend Edgar Anderson spent four years (1923–27) alone in the field tracking Northern blue flag irises across North America in order to observe as many as possible in bloom.[68] Kinsey was in good company with other contemporary taxonomists who also collected vast quantities of specimens.

Kinsey sent a copy of his study to Wheeler, who noted and praised in an enthusiastic letter the cross-species applications of "Gall Wasp Genus *Cynips*" to other insect groups and mollusks. Kinsey replied, "I have been very much pleased to find the number of people . . . who find my concept of species in accord with their own experience. This verification makes it look as if we were dealing with material well enough established to build upon in future taxonomic-genetic investigations."[69] As a result of the positive reception of "Gall Wasp Genus *Cynips*," Kinsey more energetically promoted the importance of gall wasps to scholarship in evolutionary science generally. He wrote to Henry Fairfield Osborn, of Columbia University and the American Museum of Natural History, after reading the text of Osborn's lecture at the American Association for the Advancement of Science meeting in December 1933. Osborn postulated that the order Proboscideans, including several extinct species and present-day elephants, "ranks next to man in biological importance and far surpass the mechanically inferior man in demonstration of all the main principles of biomechanical aristogenesis and alloiometry." While the scientific community never accepted aristogenesis—the idea that species evolve in anticipation of their future circumstances—what angered Kinsey was the assumption that Proboscideans were as biologically important as Osborn stated. Kinsey wrote him and asked, "I wonder if this is not an error. Would you not agree with me that the gall-wasps rank next to man in biological importance and far surpass him in demonstration of all the main principles of evolution?"[70] Osborn either did not reply or his reply has not survived, and so it is impossible to know what he thought of Kinsey placing gall wasps at the center of evolutionary studies. Regardless of what Osborn concluded after reading Kinsey's letter, the letter shows Kinsey's enthusiasm, however overexcessive, for the genus and for the possibilities it held for new and comparative scientific discovery.

Kinsey's work mirrors that of other American taxonomists at the time who were also collecting widely and splitting species narrowly. They all had similar hopes of mapping the processes of speciation, and were creating data sets that potentially could have a broader impact on the life sciences. Kinsey, along with his fellow flora and fauna collectors, were perennially on the lookout for material that could support new understandings of and insights into evolutionary theory. He began to consider in more depth how his research could contribute to contemporary arguments on evolution that concerned all life scientists. He slowly realized that large samples alone as an indicator of data accuracy would be insufficient to support his (or anyone's) theorizing about evolution. He knew that he was not alone in thinking that the question of variation's effect on evolution—along with related questions concerning the roles of hybridization, mutation, and adaptation in creating new species—should be addressed using mass gathering and breeding of specimens, close observation of data, and the formation of conclusions based on those observations.

Kinsey wrote to his graduate student advisee Ralph Voris while on a gathering trip to Kentucky in 1931, "I must have passed on to you Wheeler's advice, that the research I did in the first ten years out would determine my entire future in research."[71] And so Kinsey's first ten years out of graduate school indeed determined his scientific method for the remainder of his life: an emphasis on collecting numerous specimens, a dedication to scientific objectivity through the development of trained judgment, a focus on individual variation, an awareness of the broad scope of life science research, and an eye to linking his own findings to those of others in the academic life science community. However, a variety of shifts were underway in academe—one from field-based biology to laboratory-based biology among them—that would play a role in Kinsey's shift from insects to human sexuality.

By the mid-1930s, Kinsey had developed a sense of himself as a life scientist, a pattern of collection, a set of taxonomic methodologies, a laboratory organization structure, and a data management program that gave him usable frameworks for scientific research. He had developed systems that provided him with easy data access from which he could discern new species and publish them. As North American survey collecting neared its end, however, scientists' practices and reasons for collecting were shifting. Life scientists' questions about how best to approach the study of evolution were changing as well. Kinsey's work would change too—along with his colleagues, he would adapt to vicissitudes in the academic landscape by shifting the focus of his scrutiny and the methods he needed to approach it to match. As the library theorist Jesse Shera puts it, "Every scheme is conditioned by the intellectual environment of its age or time; . . . there is not, and can never be, a universal and permanent classification that will be all things to all men; and . . . each generation may build upon the work of its predecessors, but must create its own classification from the materials that it has at hand and in accordance with its own peculiar needs."[72] Kinsey's ability to organize, classify, synthesize, and build on disparate bodies of data—and to manipulate them to different ends, such as high school textbooks and teaching—would continue to develop as he began maneuvering into a new academic era that prioritized examining scientific processes over scientific objects.

2

The Evolution of a Taxonomist

I have kept the cynipid research going, though it has slowed because of the other research problem.
—Alfred C. Kinsey to Ralph Voris, c. October 1939

KINSEY'S SCIENTIFIC WORK in the late 1910s through the 1930s was in two broad categories: writing and teaching. As he wrote a book on edible wild plants, wrote and edited three different editions of a high school textbook and two different editions of a workbook, and published his last texts on gall wasp speciation, his academic interests began to change. His coauthored book on edible wild plants demonstrated his curiosity about wild foods and his interest in reordering and challenging accepted forms of classification. Through writing and editing his high school textbooks and workbooks, he developed a philosophy of science for himself and for young people that blended the basics of biology, including an introduction to evolution, with an enthusiasm for the application of science to everyday problems and concerns. That his textbooks and workbooks sold well throughout the country was a testament to his ability to make studying the life sciences interesting to high school teenagers and prompted him to write a biology pedagogy text for like-minded instructors as well. Through continuing his gall wasp research into the early 1940s despite challenges to his theoretical framework from other evolutionary scientists, he forged strong connections with colleagues in the United States and Europe who were intensely interested in the application of their work to human development. While he left the formal study of gall wasps behind him in the early 1940s, he took with him into the next phase of his research an abiding interest in gathering large data samples to make claims in the human sciences. Together, his writing and teaching prepared him to take on the new intellectual challenges of the marriage course and to consider a larger project on human sexual behavior.

The Use and Misuse of Classification

Edible Wild Plants of Eastern North America was the product of a collaboration between Kinsey and Merritt Lyndon Fernald, who taught botany courses at the Bussey Institution. Fernald remains best known for coauthoring the seventh edition of Asa Gray's classic *Manual of Botany* (1908). Kinsey, while completing courses at the Bussey and writing his dissertation, found time to draft a book on edible wild plants in the region of North America with which he was most familiar. Early in his three-year graduate career, in late 1916 or January 1917, he enlisted Fernald's assistance, "undoubtedly to help flesh out its technical plant treatment."[1] Kinsey appears to have finished the book before he finished his dissertation and left on his gall wasp collecting trip in 1919–20. Portions of the first forty pages and the bibliography survive in manuscript form in Kinsey's hand at the Archives of the Gray Herbarium at Harvard University. The lead author of the remaining three hundred pages remains unknown, but the text throughout maintains the scholarly yet cheerful and occasionally humorous tone of the first forty pages.

Kinsey and Fernald drafted the book between 1917 and 1919 but did not find a publisher for it until 1943, when the US army, "then in the throes of the Second World War, used the book in its wilderness survival training program." Though the first edition lost money initially, it has been revised and reprinted twice (in 1958 and 1996) and remains "among the best of its kind for the number of species it covers, the accuracy of its descriptions, and the practicality of its recommendations for harvesting and preparing wild foods."[2] Kinsey and Fernald divide their findings into fourteen chapters, according to how the nearly one thousand plant species are best prepared for consumption. Each entry contains a textual description of the plant; a short discussion of how previous authors evaluated its palatability; and how successfully the present authors prepared it. Tales of preparation failures and successes often include dry humor, such as the entry for celandine, or swallowwort: "It is a member of the Poppy Family, in which toxic or narcotic properties are frequently present, and has long been viewed with suspicion. However, when young and inexperienced domestic rabbits . . . get into a patch of *Celandine*, they devour all within reach. Taking this hint, the senior author has eaten young *Celandine*-leaves, dressed with oil and vinegar, with some enjoyment; nor has he acquired a habit from so doing. Perhaps *Che-*

lidonium has been slandered."[3] The authors' desire to upend standard rules of eating permeates the book, and suggests that they enjoyed challenging traditional expectations of human behavior.

Edible Wild Plants illustrates three themes in Kinsey's early thinking about the meanings, uses, and purposes of classification: its essential role in human survival, its use in marking racial difference in human society, and when it was and was not important. First, the basic grouping of edible plants into poisonous and nonpoisonous varieties was a vivid means of noting the importance of healthy, necessary classification: identify a poisonous mushroom as a nonpoisonous mushroom, and the eater of that mushroom could get violently sick and could die. In bold type, the authors caution readers over and over that "the beginner must be extremely cautious about eating wild mushrooms and should never allow himself to be tempted into eating any mushroom unless he is absolutely certain of its identity."[4] Proper, knowledgeable classification of wild foods meant the difference between life and death, and so careful study of the natural world and its divisions was critical. Furthermore, the chapter on mushrooms suggests the broader importance of classification as a mechanism essential to Kinsey's and Fernald's desire to make sense of the wider world.

Secondly, *Edible Wild Plants* contains Kinsey's reflections about the nature of race. The authors make repeated distinctions between "whites" or "Europeans" and other racial groups based on which wild plants each group was able and willing to eat. In the introduction, written in Kinsey's hand, Kinsey points out that "the thick soups prepared from powdered nuts and from Sunflower-seeds have been in repute among the American Indians and have been highly praised by the Europeans who have tried them." Some foods "eaten by the American Indians are so unpalatable to the European taste that until some method of preparation is found by which their undesirable qualities may be removed, they are likely to be ignored." When discussing the tastiness of four wild grains that could be used as breakfast cereals, he states, "Of these, Arrow-grass, on account of the peculiar, oily flavor of its seeds is not likely to be palatable to the European taste."[5] He argues that Europeans in the 1910s were not generally open to new flavors in their diets, but, as in the case of nut or sunflower soups, they could be pleasantly surprised.

While Kinsey may not have written the racial observations in the remainder of the book, he agreed with Fernald's enough to agree to be a coauthor of them. For example, in the entry for cat-brier, they note that "the new shoots are eaten by Indians and Negroes. We have not tried them; but if by eating every new shoot we could discourage this obstructing and fierce species, we would gladly do our share." They encourage their American readers to try some plants that others enjoyed, such as the Jerusalem artichoke: "The tubers have been in considerable repute in parts of continental Europe, but although often found in our markets, they are not greatly appreciated by the whites."[6] Here, the authors make a distinction between white and nonwhite consumers in the United States, suggesting that white Europeans were more open to, and interested in, gastronomical variety than were white Americans. White American diets were more homogenous and bland than those of other Americans.

The discussion of the castor-bean is the clearest articulation of the young scientist's and his mentor's thoughts about race: "As one reads of the plants sometimes eaten by the natives of Java, for instance, or in other regions where habit or racial differences may have established immunity he is impressed with the soundness of the advice given nearly two and a half centuries ago by John Evelyn" (*Acetaria, A Discourse of Sallets*, London, 1699). They list some plants that, according to Evelyn, natives can eat that Europeans cannot, and conclude, "When we read of oriental people cooking and eating the plant of Castor-bean or even eating the cooked seed, three of which raw, the source of castor-oil, would kill an ordinary occidental, it is evident that racial differences extend beyond color, speech and methods of thought."[7] For Kinsey and Fernald, racial difference manifested in multiple ways: in skin color, language, and thought patterns. Racial difference also developed as part of cultural conditioning. Racial differences extended at least to the ability to enjoy and to digest certain foods, and to the cultural acceptability of eating them. Race was at once a physical, mental, and cultural concept. However, in all of the mentions of racial difference here and in the dozen or so more instances scattered throughout the book, racial difference exists but has no hierarchical connotation. It marks differences among groups, but no one group is better than any other. The racial divisions on the sex history interview form Kinsey developed in the late 1930s include nine possible classifications of race, and the notation on the question indicates that Kinsey was interested particularly in how cultural expressions of racial identity related to sexual behavior. The idea that psychological conditioning was an important part of a person's sexual makeup would form a critical element of the argument about male–female difference in the *Female* volume.[8] If Kinsey had written further volumes on his African American, Native American, and Asian American histories, he may have been able to explore further his ideas of the interrelationship of race and sexual behavior.

Thirdly, unhelpful classifications are a prominent theme in *Edible Wild Plants*, particularly in the introduction. In the very first sentence, Kinsey points to the useless differences that many people have drawn between "civilization" and "primitive life": "Nearly everyone has a certain amount of the pagan or gypsy in his nature and occasionally finds satisfaction in living for a time as a primitive man." Crossing an artificial barrier between domestic life and life in the wild had the potential to bring people much pleasure. Furthermore, breaking down the boundary between wild and civilized food was a means of breaking down prejudices as well: "In a highly 'civilized' community we are so used to the conventional dishes that there are some among us who have a prejudice or squeamishness about eating 'weeds.' . . . [T]hese common weeds make wholesome and really delicious food, when properly prepared, and the prejudice against them is chiefly due to the unsavory connotation of their names."[9] "Weed" as opposed to "edible plant" was a pointless construct. Kinsey's interest in articulating accurate and clear classification systems while dismantling meaningless classification systems would form an important part of the *Male* volume's criticism and reclassification of human sexual identities.[10]

Kinsey and Fernald clearly connected misclassification with unhealthiness, and proper classification with health. It would not be wise to push Kinsey and Fernald's calls for tolerance too far, into a call for racial and sexual equality for all whose time had not yet come. Nonetheless, it is remarkable that they argued that human difference, whatever its physiological or cultural groundings, need not be hierarchical, in a Progressive Era American society rife with legal and social prejudices against anyone nonwhite and nonheterosexual. *Edible Wild Plants* showed that classification was significant for human understanding, food, survival, and satisfaction, but if misapplied, it could be pointless and harmful. The wise scientist knew the difference, and said so.

Developing a Philosophy of Science: Textbooks and Workbooks

Kinsey's biology lectures on secondary education, textbooks and workbooks, along with his published articles on teaching, show his engagement with evolutionary questions and his willingness to teach evolution to his students. Kinsey emphasizes in his written life science materials the necessity of students observing the natural world firsthand in order to learn about it and to understand it. Kinsey exhibits his scientific methods in his textbooks, workbooks, and biology teaching guide, and they reveal that his interest in studying human sexuality developed at least partially out of his attention to it as an educational problem for himself and for secondary science teachers. He did not believe that human sexuality and reproduction could be left out of a synthetic overview of life science, any more than evolution could, as young people needed a thorough scientific education in order to become fully knowledgeable citizens.

Kinsey's textbooks, workbooks, and lectures on biology teaching in high schools show his engagement with evolutionary questions and his willingness to teach evolutionary principles to his students. Above all, Kinsey emphasizes in his textbooks and lectures the necessity of training students to observe the natural world carefully and precisely in order to understand it and to make good life decisions as they move through it. Kinsey encourages students to learn techniques of observation and scientific method in all of his secondary educational material. In a letter to Fernald, he acknowledges the boldness of attempting such a thorough synthesis of life and human sciences in a single textbook: "I realize the stupendous audacity of attempting to cover as diverse fields as I treated in this volume, but somebody has it to do for the beginning student or else he will get such one-sided training as most of our schools have contributed up to date."[11]

Kinsey was the first and one of the only American college biology professors to write a high school textbook in the first half of the twentieth century. His *Introduction to Biology* was not the most popular high school biology textbook of the 1920s and 1930s, but it sold well because he wrote it in an engaging and accessible style and filled it with encouragement to young people to explore the wider world. Kinsey's biographers suggest that he wrote the textbooks and workbooks to earn extra money for his family, given the low salaries for university professors at public institutions in that era, though he also seems to have been genuinely interested in improving science education for American high school students.[12] Through his textbooks, workbooks, and lectures, Kinsey advocated for practical science education that introduced students to the basics of scientific method, to the importance of naked-eye observation, and to an ecological worldview. Edgar Anderson linked Kinsey's textbooks and workbooks and his teaching directly:

> He organized and taught a general course in biology meant particularly to train teachers of high school biology for the state of Indiana. Within a few years this resulted in a high school text in biology so excellent and so original that some of its chapters are still [in 1961] assigned as background reading for graduate students in biology. With his wide knowledge of both plants and animals and his years of experience in the field with boys and girls it was one of the best, as well as one of the most successful biology texts which has yet been written. It was accompanied by a laboratory "Work Book" which demonstrated for the thoughtful teacher how some of the most fundamental problems in biology could be demonstrated and even experimented with in a simple sort of way, on almost any vacant lot in the city or the suburbs. One did not need fancy equipment to teach biology, the science of plants and animals. Plants and animals of some sort are always about us in this world and are always doing interesting things. We can easily study the ones at hand.[13]

Anderson, who knew Kinsey from graduate school until the latter's death, saw the interconnections between Kinsey's different academic interests in hindsight. Kinsey gave his own perspective on what an ideal biology course should and should not contain in an address to Michigan schoolteachers in 1929: "Instead of utilizing the living things in the pond or roadside ditch, we have been taking a microscopic bit of pond scum, magnifying it beyond all familiarity, compressing it out of all third dimension, inverting it until it looks dizzy, spending an hour convincing the student that the real world must be observed through lenses, and spending the rest of the week developing his ability to make artistic reproductions of this far-fetched sample of reality."[14] Kinsey's view of the content of ideal biology courses instead emphasized student examination of the natural world through their own eyes, not only in a detached manner through scientific instruments. Furthermore, many other textbooks confined discussions of reproduction to flowers, insects, and frogs, which suggested to Kinsey that "the more important purpose was to make discussions of sex and reproduction as unpalatable as possible."[15] For Kinsey, biology instruction needed to focus on the basic goal of developing in teenagers an interest in the world in which they lived and the ability to use scientific methods for interpreting that world. Only when students understood basic biological principles, including their own sexual and reproductive processes, and scientific method, should instructors expose them to applied science. Kinsey believed that starting students' scientific learning with applied sciences like economic entomology or agricultural science—practical forms of science popular in a rural agricultural state like Indiana—robbed them of a broader comprehension of the processes of the natural world and of the basic knowledge they needed to support applied methods. The idea that basic life science was necessary before applying it to immediate problems would also shape Kinsey's approach to studying and writing about the processes and contexts of human sexual behavior.

Kinsey likewise emphasized students' need to learn biology as a holistic science, rather than separating the plant and animal worlds for pedagogical convenience. In his view, life sciences at the secondary level needed to show students how plant and animal worlds worked in tandem, stressing "the principles which are common to both worlds," as students themselves lived and functioned in a world in which plants and animals interacted together.[16] They needed to understand the natural world as it actually functioned. A basic, thematic, and practical view of the natural world was most usable for the future American citizens in high school biology classes, in Kinsey's thinking, as few of them would need the specialized knowledge and research techniques of professional biologists. Having students memorize lists of species or making them observe teacher demonstrations without being able to experiment themselves was a waste of time when students could be learning about biological processes that concerned all types of creatures. Kinsey wrote in 1926 for high school instructors in Indiana, "the principles of heredity are neither botany nor zoology; the principles of evolution are broadly biologic . . . and in elementary teaching can be presented best from both plant and animal data."[17] Kinsey demonstrated his commitment to teaching scientific processes over memorizing data objects throughout his high school pedagogy materials.

Kinsey's three high school biology textbooks (published in 1926, 1933, and

1938), two high school workbooks (published in 1927 and 1934), and guide for teaching high school biology (1937) show his commitment to ensuring the usefulness of biology instruction to contemporary students. They exemplify his ability to handle difficult subjects, and the fact that he was conversant in the processes of synthesizing life science research. His textbooks and workbooks cover aspects of biological science from protoplasm to "Mountain-Top Biology," from birds to "The Value of Scientific Research." The textbook's preface also lists thirty people who read the book manuscript to offer feedback and advice, from the IU professor Paul Weatherwax (who reviewed the botany sections) to the Princeton professor Edmund Conklin (who reviewed the evolution and heredity sections).[18] Kinsey's interdisciplinarity and commitment to scientific method, along with the impact of his graduate adviser, are all evident from the opening pages of his textbook.

In 1926, Kinsey sent a copy of his first textbook to his doctoral adviser William Morton Wheeler with a letter and an acknowledgment of Wheeler's thinking on his own work. "I find . . . that I owe a great deal to you and your continued treatment of questions in a broad, biological way."[19] Wheeler's response is not extant, but if he read the textbook, he could see clearly how he shaped Kinsey's approach to teaching life sciences. For example, Kinsey defines scientific method in the first edition of his textbook along the lines that Wheeler outlined as "a scientific skepticism, an emphasis on the importance of direct observation, a wary evaluation of authority, and an interest in discovering the causes of observed phenomena."[20] Kinsey maintained some form of scientific method rooted in that definition throughout his life, and then taught it to a generation of high school and college students through his textbooks, workbooks, and teaching. That definition of scientific method would shape his approach to sex research as well.

Kinsey's high school textbooks cover numerous subtopics of biology in an accessible style and include a few pages on evolution. For example, the section "Further Evidences of Change," contains a delicate discussion of evolution, deemphasizing the role Charles Darwin played in the history of science and calling evolution "change." As Ronald Ladouceur puts it, "in the years following the Scopes Trial, authors and publishers found that a few simple linguistic tricks were all that were necessary to keep community objections to the adoption of their textbooks to a minimum."[21]

Nonetheless, student users of the textbooks were introduced to some of the basic concepts of evolution: hybridization, selection, and mutation. While many high school textbooks proceeded through the plant and animal world "up" an evolutionary chain, "Kinsey favored a unit structure based on key subdisciplines—morphology, physiology, genetics, ecology, [and] distributional biology." The unit on behavior is the conclusion, as "the study of what organisms do, and why they do it, may be made a very exciting climax to biology." The closest Kinsey comes to discussing human sexuality is in the 1933 edition's "Reproduction" chapter: "Among other higher mammals, including the insects, reptiles, birds, and mammals, the sperm are placed directly into the body of the female, where they have still better chances of meeting the eggs."[22] He does not mention how sperm actually reached the egg inside of the body of a female animal, however. Kinsey discusses human reproduction in the textbook as just another kind of mammalian reproduction, deliberately not drawing controversy or any special attention to the subject but not ignoring it either. Unlike other biology textbooks that emphasized that humans should improve nature for their own benefit, Kinsey's textbook displaced humans as the center of the natural world, promoting a holistic "unity of life" approach that matched his focus on introducing students to a comprehensive view of the natural world obtained through observation.[23]

Kinsey designed the workbooks that accompanied the textbooks to put his commitment to teaching observation into practice. He encouraged students to keep a well-organized and clearly written record of all of their observations, as he had done for his gall wasp records and as he would do for his sex histories. In the 1927 edition of the workbook, *Field and Laboratory Manual in Biology*, he also directs instructors to ensure that students are schooled in scientific methods, not merely facts: "The scientist's faith in sensory observations as his fundamental source of knowledge, an attitude of wary acceptance of second-hand knowledge (authority), a consuming urge to search for the explanations, the *reasons*, the *causes* of phenomena, are the most practical things our students may acquire." He tells teachers and students that they should trust themselves and their own observations as the most reliable sources for discovering the realities and truths of the natural world. As he wrote in the 1934 edition of the workbook, "You are asked only to use your eyes and the rest of your sense organs, to see what you can see, to believe what you see, and to question every other man's word when it does not accord with your own observations. This unlimited dependence on what we observe is what we call science."[24] As Kinsey taught it, observation itself was science, and teachers and students only needed to look closely at the world around them—and take good notes—to practice it. Many of the activities in both editions of the workbooks involve simple trips to a nearby grocery store, dairy, butcher shop, farm, park, or even an urban weed lot so students could describe ecological processes (such as mold growing on old produce, the use of manure for fertilizer, or squirrels storing nuts in trees). Further, as Kinsey revised his workbook for sale during the Depression, teachers and students were likely heartened that they could learn and practice science at its best at little to no cost and by using a minimum of special equipment, tools, and field trips.

The textbooks' sales records show their acceptance by and accessibility to American high school students. They sold the most copies in the South in the late 1920s and early 1930s.[25] (It was also useful in a state college setting, as Kinsey's former graduate student, Ralph Voris, used it in courses for biology secondary education students at Southwest Missouri State Teacher's College.) The primary school districts that bought the most textbooks and workbooks were in Florida, Texas, Alabama, North Carolina, Kentucky, Arkansas, and California. Thus students whose parents were hostile to teaching evolution in schools were being introduced to evolution anyway. Neither they nor their parents may have recognized that evolution was part of their science curriculum. Kinsey had no telltale ape-to-man illustrations but included photographs of mutations in blueberries, bees, daisies, pigeons, dogs, and ferns.[26] At least in some areas of the South, Kinsey's publisher was successful in selling textbooks that included instruction on evolution to public schools whose officials were undoubtedly attuned to its representations. Thus Kinsey was able to include evolutionary concepts in his textbooks by modifying the language that he used to discuss them and the images that he used to represent them. The first textbook sold 118,897 copies; the two subsequent editions together sold 109,976 copies; *Methods in Biology*, Kinsey's biology teaching guide, sold 1,929 copies; and the two editions of the workbooks together sold 50,687 copies. Kinsey's pride is clear when he told Edgar Anderson, "Allowing for the use of each book by several students, we have talked to more than two thirds of a million boys and girls through these books."[27] Kinsey's success with textbooks perhaps inspired the accompanying teacher's handbook and his first published explication of his views on teaching human sexual behavior.

Kinsey considered revising the textbook one last time in 1942 but ultimately decided against it, likely due to the time constraints that the work on his biology courses and the sex history interviews were then imposing on him. (He gave up textbook revision and the gall wasp work at the same time.) He contemplated minor changes like changing chapter titles along with a more extensive topical restructuring due to students' immediate need to learn the effects of war on the life and human sciences. He ruminated that students of biology in wartime needed to know "the effect of the war on agriculture in having to feed this country and our allies; community health measures designed to protect the civilian population in the event of air raids and the bombing of large cities; some specific helps on planting and growing 'victory gardens'; the more-than-usual necessity for the average person keeping physically fit by personal healthful habits instead of relying on the advice and care of doctors who are needed in the armed services—in other words, how the biologists and a knowledge of biology are helping to win the war."[28] Kinsey clearly viewed biology as a holistic science with applications across the war effort. He also advocated the importance of young people's knowledge of science for being informed citizens on the home front. Like other textbook writers at the time, Kinsey emphasized in his scientific method the personal characteristics young people needed to enact it successfully: curiosity, creativity, independence, prudence, worldliness, and industry—all popular middle-class American values.[29] While Kinsey never made a fourth revision of his high school textbook, his ideas for a fourth edition show his continuing interest in linking young people's scientific education to their ability to participate fully in American civic life. So as he finished the third edition of the textbook, he decided to share his pedagogical methods with fellow biology instructors.

Methods in Biology, Kinsey's 1937 biology teaching guide, did not sell even one-twentieth the amount of workbooks or textbooks, and it had nowhere near the same number of readers. Regardless of the number sold, though, of all Kinsey's teaching writings it most clearly showcases Kinsey's philosophy of science. The text of *Methods in Biology* suggests that he had given considerable thought to the problems of teaching human sexuality—including masturbation, venereal disease, and reproduction—to young people well before his semipublic discussions of human sexuality in the Indiana University marriage course in summer 1938. Kinsey states in *Methods in Biology* that animal behavior broadly, and sexual and asexual reproduction in particular, are so central to teaching biology that it is almost impossible to teach biology without them. Parents, he thought, would not object if their children were being taught "the strictly biologic phenomena"; but too often high school teachers went "beyond the biology into the social and moral problems involved." Some teachers, Kinsey found, considered it their duty to cover "the personal and social aspects of sex in the course on biology." Kinsey believed that biology teachers were rarely equipped to handle the nuances of evenhanded ethical evaluations of different sexual behaviors. They also were prone to focusing on abnormalities or venereal diseases, or to condemning nonreproductive sexual behaviors like masturbation. A focus on those topics by the embarrassed, moralistic, or inexperienced biology teacher, according to Kinsey, could do "a great deal more damage than good."[30]

Kinsey suggested instead that biology educators should read Katharine Bement Davis, Robert Latou Dickinson and Lura Beam, Gilbert V. Hamilton, and William S. Taylor for ideas on teaching human sexuality. He had been familiar with their works since at least April 1935, when he presented on them at an IU faculty research seminar.[31] Each of those authors conducted and then published the results of written and oral interviews on sexual behavior. While Kinsey disagreed with some of their findings, he applauded their attempts to gather behavioral data without religious or moralistic overtones, arguing that if sex and reproduction were taught, as these other scholars had researched, without "any reference to the social and moral problems involved, and above all avoiding any emotional display in the presentation of the material, the reactions of the students should present no difficulties." Kinsey had also identified a more serious problem underlying poor sex education. "Under the guise of science," he continues, "we too often have sex instruction which is a curious even if a well-intentioned mixture of superstition, religious evaluation, and a mere perpetuation of social custom."[32] Kinsey thought it best to use a version of scientific method based on observation for evaluating materials to teach high school students about biology and human reproduction. Further, Kinsey took his working scientific method, based on observation and integrative data management, to the study of human sexuality, rather than learn new laboratory-centered methods of studying evolution. Wheeler had introduced him to the possibility of applying scientific methods to human problems in graduate school. As Kinsey included sex education in his own course for secondary education biology teachers and discussed the topic in detail in *Methods in Biology*, it is clear that the problem of teaching human sexuality was one of his long-term intellectual interests.

Though *Methods in Biology* probably had little impact on life science teaching in the United States, it is valuable for understanding the development of Kinsey's views on sex education. He was familiar with five books that would have notable impacts on the formation of his sex history interview and the ordering, explanation, and comparison of data in the *Male* and *Female* volumes. Kinsey often critiqued previous research as a means of forming his own ideas about a subject. In *Methods in Biology* he states that "with many of the conclusions in [Maurice Bigelow's *Sex-Education*] the recent objective studies are not in accord," and that "with the general attitude in [John F. W. Meagher and Smith Ely Jelliffe's *A Study of Masturbation and the Psychosexual Life*], and with many of the specific conclusions, the recent objective studies are not in accord."[33] He states earlier in the book one of the premises for his approach to sex research, which he echoed repeatedly in his printed work and letters: "Ethical evaluations, questions of good and bad, of better or worse, of right and wrong, are, like determinations of beauty, outside science."[34] Bigelow's and Meagher's books were overly based in philosophy for his taste.

Kinsey's short but pointed rejection of Bigelow's and Meagher's works on sex education and masturbation, respectively, was just one example of how his evaluative processes worked. He aligned his own thinking with the more "objective studies" and positioned himself as an "objective" critic of their writing. He disliked Bigelow's work for its focus on improving sexual behavior only as a means of improving heterosexual domestic life, its promotion of eugenics to govern human reproduction, and its hostility to homosexuality and to anyone teaching sex education "who lack[s] thorough physiological training and whose own sexual disturbances have led them to devour omnivorously and unscientifically the psychopathological literature of sex by such authors as Havelock Ellis, Krafft-Ebing, and Freud."[35] Meagher and Jelliffe viewed masturbation as nearly universal among humans, but also as an infantile way of expressing sexuality that they roundly rejected as appropriate for adults. Through Kinsey's objections to authors like Bigelow, Meagher, and Jelliffe, his thoughts on the best ways to teach sex education began to take shape: the need to discuss controversial subjects evenhandedly; favoring works based on observation; avoiding traditional ethical, religious, and moral judgments; and intensive reading of all literature in the field, good or bad. Kinsey may also have had in mind the religiously influenced works of his nemesis on the IU School of Medicine faculty, Thurman B. Rice, whose shaky authority on marriage for collegians Kinsey would challenge with his lectures in the IU marriage course in spring 1938.[36]

Kinsey's involvement in shaping high school biology education did not end with sneaking evolution into his textbooks and workbooks, publishing well-received synthetic textbooks, and authoring a book on pedagogical philosophy. Archival records of his participation in Indiana and in nationwide high school biology curriculum development from 1938 to 1943 show his interest in fostering scientific knowledge and methodologies among young people and a tolerance for diversity. One of the goals listed in "Preliminary Statement: Biology in the High School Curriculum" from a committee (including Kinsey) working to improve science education in Indiana high schools was "to help the student develop an adequate social philosophy. . . . [That] would involve an increased tolerance of the wide variation in people and behavior; an intelligent attitude toward race relations; [and] a functional understanding of the evolutionary viewpoint." One aspect of Kinsey's fostering of tolerance, as he wrote in *Methods in Biology*, was to approach teaching human sexuality without sensationalism or embarrassment on the instructor's part. Perhaps he had his first sex history interviewees in mind as he advised those committees. Teaching students tolerance in their high school lives could help prevent emotional difficulties later in life, he thought.[37] Mentioning the need for "an intelligent attitude toward race relations" indicated his opinion that racial attitudes among students needed improvement. Kinsey's biology-based ideas of proper sex education for young people extended to his view of studying sexuality. At the same time that Kinsey's ideas on textbook writing were beginning to change, his ideas on evolutionary theory and the intellectual context for them was changing as well. The years 1936 through 1938 were critical to a shift in Kinsey's focus from life sciences toward human sciences.

The Origin of Higher Categories in Cynips *and the Evolutionary Synthesis*

The ideas on evolution in Kinsey's 1936 book, *The Origin of Higher Categories in* Cynips, would place him at odds with some members of the evolutionary science community, and the book's reception would prompt him to rethink his place in the life sciences.[38] When the Society for the Study of Speciation

(SSS)—the primary society for evolutionary study in the United States—was founded in 1940, however, Kinsey participated in the organizational meeting. The synthetic works in evolutionary studies throughout the 1930s and 1940s were among his various inspirations to create his own synthesis of research techniques, taxonomic practices, and theories. Many of his scientific peers were motivated to do the same.

Kinsey's last book-length study was published just as the intellectual movement known as the evolutionary synthesis was taking shape. What the synthesis actually was, however, remains a vexing historiographical problem. As Joe Cain, one of the foremost scholars of the synthesis, has stated, "we can't agree on whether this was an intellectual activity, a social one, or a political coup. Even for single sources, the instability of meaning serves as a signal that polyvalence is at work and a different conceptualisation of the problem is required." Cain identifies four organizing threads capturing evolutionary studies work in the synthesis period: (1) the nature of species and the process of speciation; (2) four primary problem complexes: variation, divergence, isolation, and selection; (3) methodological changes in experimental taxonomy; and (4) the ways that each of those threads coincided with larger trends, "especially a shifting balance in the life sciences towards process-based biologies and away from object-based naturalist disciplines."[39] The evolutionary synthesis was a long-term process that included numerous intellectual exchanges across multiple fields, and such exchanges shaped both the synthesis and the contours of academic communities around it. Studying Kinsey's *Origin of Higher Categories in* Cynips in the context of the evolutionary synthesis supports the recent scholarly contentions that it was a dynamic period of intellectual history that had reverberations across twentieth-century academe, including toward Richard Goldschmidt's "hopeful monsters," Edgar Anderson's evolutionary mapping of iris and corn hybrids, Aleksandr Promptov's theory of speciation through ornithological research, and Kinsey's own interdisciplinary sex research.[40]

Origin of Higher Categories in Cynips built on the work Kinsey had published in "Gall Wasp Genus *Cynips*" by adding sixty-nine additional species in the genus *Cynips* to the ninety-five he had identified seven years before. That addition brought the total identified species in the genus to 164. His book was based on the roughly 7.5 million specimens that he had collected, acquired, and kept his assistants busy sorting and storing through early 1936. As one historian of science has put it, "Kinsey's massive biogeographic studies of North American gall wasps were the single best source of data on variation and speciation in natural populations before [Theodosius] Dobzhansky and [Ernst] Mayr."[41] Kinsey was clearly proud of himself for working very hard and collecting so many gall wasps. His excitement at producing this major gall wasp study was evident in a September 1934 letter to Ralph Voris:

Have spent the whole summer on "Cynips and the Origin of Higher Categories." Been headed toward it since the Cynips monograph in 1930. It includes all additional data since 1930 on the genus—have over 50 new species to add to the 93 described in 1930. Mostly Mexican, but a number from our 1929 Western trip. Mexico a perfect gold mine! Two-thirds to three-quarters of my specimens bear red labels [indicating new species]! Will easily have 700 n[ew] sp[ecies] from the Mexican trip when it is all described. And the complexes and subgenera, and even one genus which is far enough away from Cynips in the UNITED STATES—all run together in southern Mexico until it can be established, I think, that the higher categories are nothing but sections in a continuous chain of species—the divisions between the categories purely arbitrary unless one has limited his collecting to an incomplete portion of the range, or unless nature has obligingly exterminated some of the older species. There is no tree of life—the simile should be to the creeping vine or plant with runners. The first species of the new genus is as closely related to the last species of the old, as any two species are to the other within the genus. Higher categories arise merely as species—by mutation or hybridization and subsequent isolation—with no greater degree of difference than is involved in the origin of any other species.

Amen.

Now, Mr. Man—tell me what your staphylinids [beetles] say to all that. I am more interested in learning whether cynipid generalizations apply to other groups than you may realize. And especially to the groups which have had some contact with modern taxonomy & modern taxonomists.[42]

In his letter, he explains to Voris his ideas about how evolution worked and wonders if Voris's research on beetles also upheld his thinking. Kinsey was enthusiastic about his work and about sharing it with the evolutionary science research community, including his former graduate student. He wanted to make a widely felt impact on that community, and thought that his intensive work on cynipids was the way to do so.

Before launching into the taxonomic findings based in such a vast compendium of entomological data, Kinsey presents three main arguments in the introduction to *Origin of Higher Categories*. The first is for a reconfiguration of the two-dimensional, unidirectional Darwinian tree of life into "an infrequently dividing chain in which the oldest species may remain coexistent with all of the derived species." That idea correlates with his description of the many new gall wasp species that follows the book's main text, which further subdivides the species he had named into subspecies or varieties. Minor subdivisions like those were common for taxonomists sorting incomplete collections, who used them to designate species in geographical ranges that did not overlap.[43] Second, Kinsey advocates a labor-intensive combination of genetics and taxonomy, bringing the best aspects of both fields together so that they reveal evolutionary data not visible separately. But his preference for field science remains clear as he criticizes lab-centered scientists whose "assumption of ultra-laboratory forces represents a spiritualism which seems out of place in science."[44] Kinsey connects laboratory-oriented science with "spiritualism," implying that laboratory science was based more in personal belief, while field naturalism was based on reality and facts. Kinsey's unworkable and unrealized vision of a genetics/taxonomy blend involved the creation of a secure domed enclosure around a natural habitat that scientists would study for changes in reproductive patterns in plants and animals over several years.

Third, Kinsey presents his thesis about higher categories of species. While he thought that species exist in nature, he asserts that higher categories—any taxonomic unit above the individual—do not. He argues that scientists could determine those phylogenetic chains through intense and wide-ranging scrutiny of existing species, though paleontologists in particular were guilty of stretching nonexistent links between ancient and modern species. He encourages his fellow scientists to focus on naked-eye observation to determine evolutionary patterns, not on theoretical linkages between extinct species back through time. Kinsey's disbelief in higher categories was typical of taxonomists at the time who were also describing species in collections they believed to be incomplete, as he still thought his collection was, despite its size.[45] However, his rejection of the usefulness of paleontology for understanding higher categories placed him in a minority position vis-à-vis those seeking a theory of evolution that would ideally and ultimately encompass all species across time.

Kinsey presented his ideas on higher categories in the paper "Supra-specific Variation in Nature and in Classification from the View-Point of Zoology" at the December 1936 American Society of Naturalists meeting in Atlantic City, New Jersey. He was on a panel with evolutionary thinkers from other fields who were also investigating supraspecific variation, or the scientific validity of higher categories (taxonomic categories larger than species). His graduate school friend Edgar Anderson also presented on the panel. Reading *Origin of Higher Categories in* Cynips in preparation for the Atlantic City ASN meeting, Anderson wrote to Kinsey, "Just now I am deep in your Olympian thunderings in regards to Cynips. With most of what you say I am in hearty agreement, and most of my disagreements are really in the nature of qualifying phrases."[46]

George Gaylord Simpson had more specific and greater disagreements. Simpson, who was a curator in the Department of Vertebrate Paleontology at the American Museum of Natural History, was on the supraspecific variation panel with Kinsey and held an opinion opposite to Anderson's. As Simpson prepared his panel presentation, he sent a draft to William K. Gregory, the chair

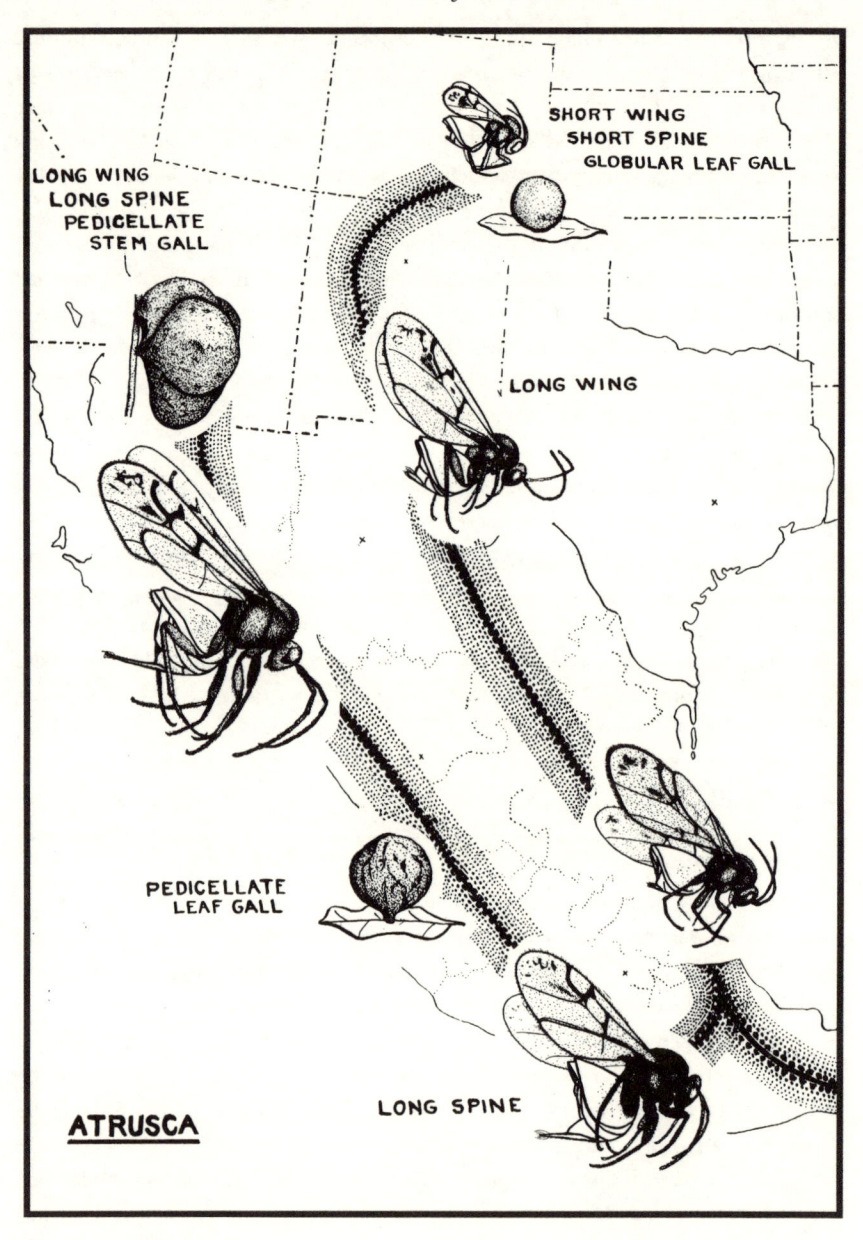

Figure 2.1. Alfred Kinsey's rendition of his "higher categories" concept using gall wasps and gall specimens in the genus *Atrusca*. "Higher Categories Are Sections in Chains of Species: Summary Map Showing Origin of Diverse Complexes in *Atrusca*." Alfred C. Kinsey, *The Origin of Higher Categories in* Cynips (Bloomington: Indiana University Publications, 1936), 4.

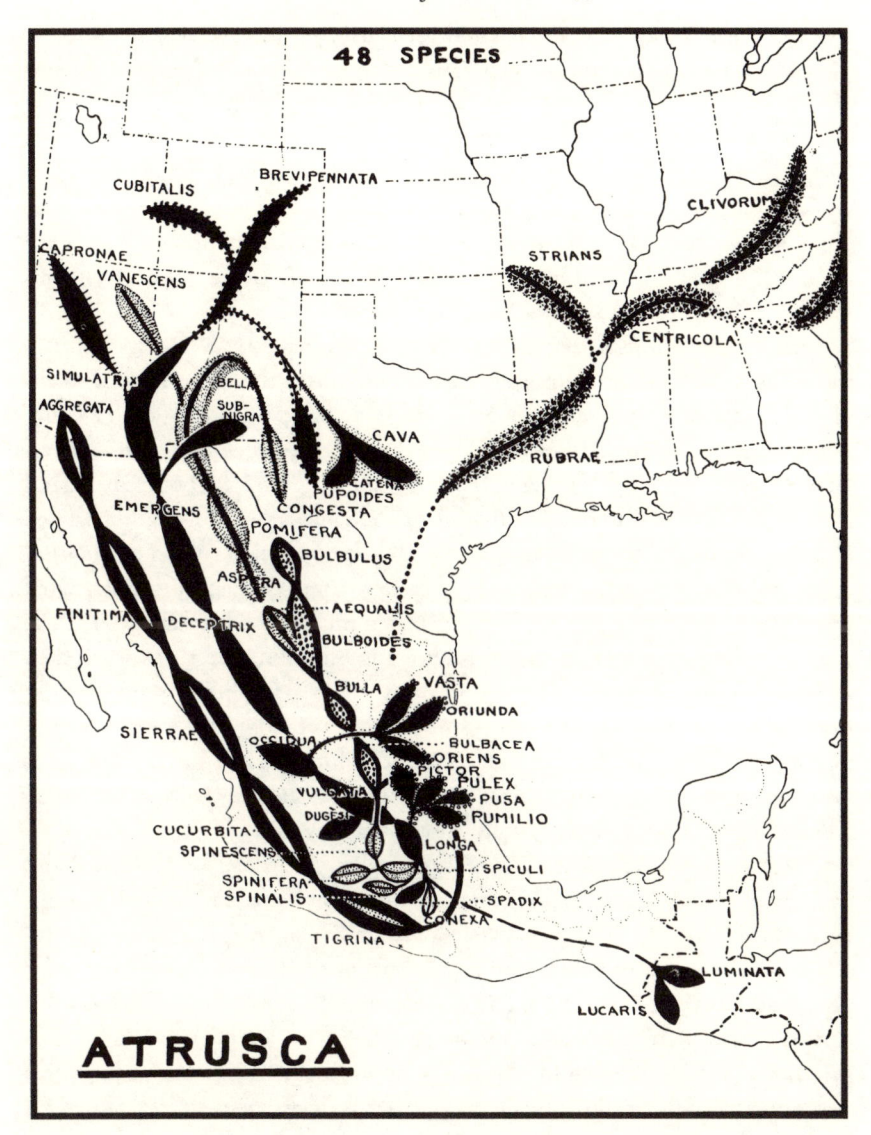

Figure 2.2. Alfred Kinsey's rendition of his "higher categories" concept using related species in the genus *Atrusca*. "Phylogenetic Map of *Atrusca*: The Diverse Complexes Are Arbitrarily Delimited Ends or Sections in a Phylogenetically Continuous Chain of Existent Species." Kinsey, *Origin of Higher Categories in* Cynips, 44.

of the panel, stating that he "had great difficulty in selecting and organizing [material], from the embarrassingly vast amount of data on hand."[47] For Simpson, Kinsey's work was so poor that he had plenty to criticize, and he used

Kinsey's entire body of work as a foil for clarifying his own ideas on higher categories and evolution more generally. "Kinsey's review of [higher categories] is the most recent and in many respects most complete, and it is based on a remarkably thorough and profound study of an exceptionally large mass of data," Simpson declared. He restated thirteen concepts underlying zoological works on higher categories that Kinsey had tried to refute, and disagreed with the biologist on nearly every one.[48] While Simpson initially praised Kinsey for his large sample size, Kinsey's dependence on sample size alone to prove his views was insufficient to garner Simpson's support.[49]

One by one, Simpson demolished the concepts that Kinsey put forth in *Origin of Higher Categories in* Cynips. For example, Kinsey thought ancestral stocks of gall wasps were not extinct; Simpson said that paleontological evidence showed that they were. Kinsey argued that higher categories did not exist in objective reality. Simpson countered that they did, as groups of species in general are more alike than different, both in the present and across time. Kinsey stated that in order to group species across time, they had to retain similar or identical characteristics; Simpson disagreed—of course, some characteristics of related species cannot be found in the present due to mutation and adaptation.[50] The entomologist Richard Goldschmidt, following Simpson, also sharply attacked *Origin of Higher Categories in* Cynips in print. Years later, Ernst Mayr thought that Kinsey handled his gall wasps "in a rather typological manner" instead of understanding "the population approach," which regarded every species as a varying population of interbreeding individuals. In the 1980s, Stephen Jay Gould called Kinsey's classifications "bloated taxonomies."[51]

There is no record of how Kinsey reacted to Simpson's or Goldschmidt's challenges, but their remarks became part of Kinsey's gradual decision to leave evolutionary studies behind. A brief exchange in January–February 1938 between Kinsey and Simpson regarding reprints of their respective lectures suggests that they parted on cordial terms. Although Kinsey boasted of the positive reception that he received from the other panelists to Ralph Voris and to Edgar Anderson and received his friends' praise in turn, Kinsey began to realize during 1937 that his arguments and data-gathering techniques were ill-suited for the evolutionary questions that his contemporaries were starting to address. As Robert Kohler puts it, "survey collecting was meant to (and did) make species taxonomy a more rigorous science, but in doing so, it ultimately transformed classification from an end in itself to an instrument of evolutionary theory." Joe Cain has stated simply that "Kinsey simply could not divorce his taxonomic decisions from larger narratives about biological processes."[52] As North American specimen collections became nearly complete through the end of the 1930s, and as more scientists turned to laboratory-created animals to investigate patterns of

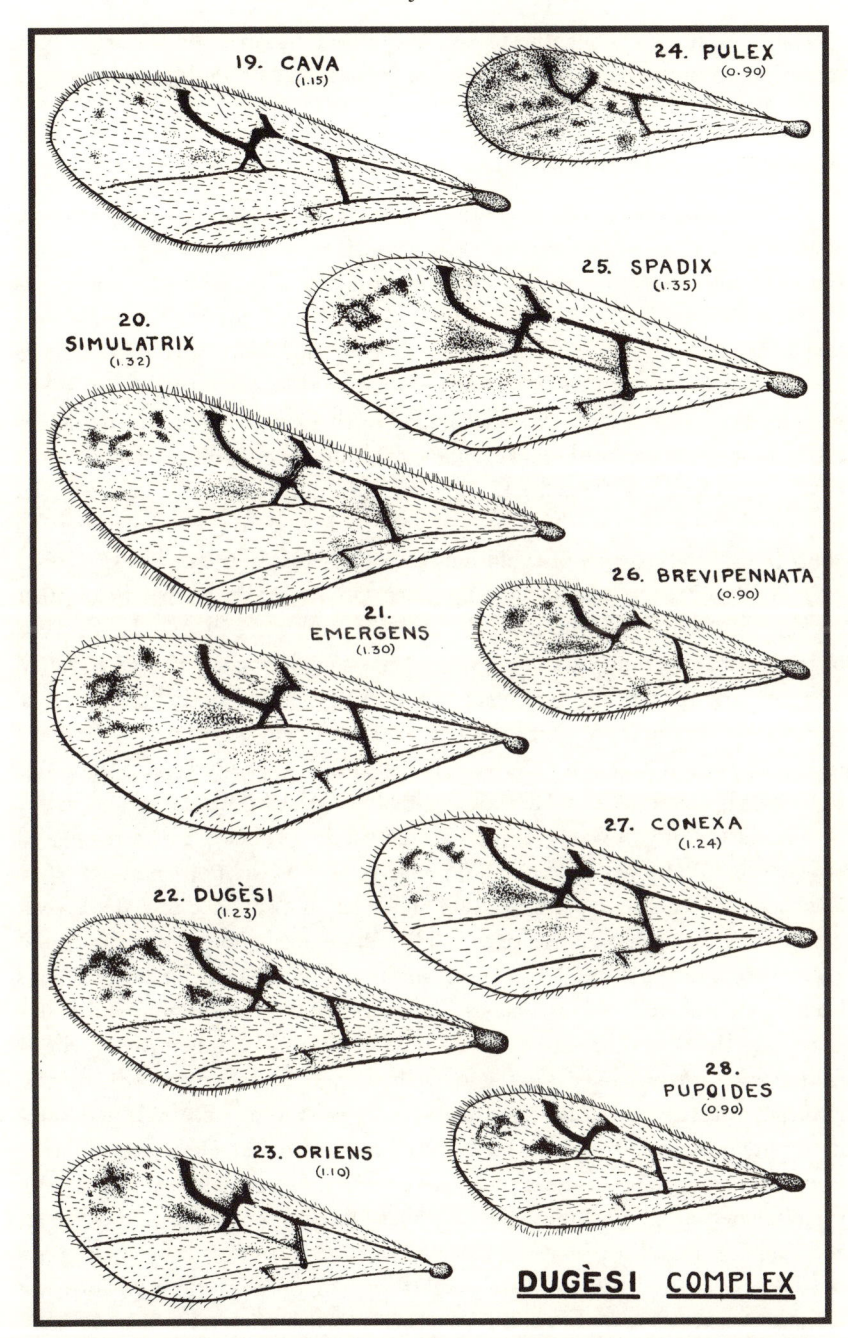

Figure 2.3. "Mutation in Wing Length in the Dugèsi Complex." Kinsey, *Origin of Higher Categories in* Cynips, 84.

speciation, Kinsey's field gathering skills and data organization patterns began to decline in importance in academic life science. But he could see how useful those skills could become if he applied them to the human sciences.

A New Experimental Animal

At the same time that Simpson and Goldschmidt derided Kinsey's assertions about higher categories, there was also a transformation in the type of animal that was most useful for professional evolutionary scholarship. Not only was the type of evolutionary research changing in the late 1930s, the type of animal that scientists used to answer questions about evolutionary processes was also changing. Kinsey had chosen to study the genus *Cynips* at a time when it had been understudied and poorly collected in North America. He selected that genus in part so that he could make a name for himself in naming and describing new species, and thus leverage his knowledge into academic recognition and improve his standing in the academic world. He had also wanted to make a contribution to the study of evolution through the use of the gall wasps. However, "each organism provides an imperfect window on the properties of other organisms."[53] Kinsey's gall wasps, like Wheeler's ants, were ill-suited to the type of laboratory breeding that the new systematic evolutionary biology required of its specimens.[54] Thus Kinsey would not only need a new study organism if he were to continue in evolutionary biology past the late 1930s. He would also need to adopt new techniques for studying it and to develop new research questions to frame his investigations. He chose a different path instead.

The nature of gall wasps themselves and their reproductive patterns played a clear role in Kinsey's decision to shift fields, as they did not lend themselves to genetic laboratory experimentation. They hatch only once or twice a year, if at all, and die after breeding once in a few days. Sometimes the wasps cannot chew their way out of the gall and die inside it. The wasps can also die if the galls become too infected with parasites or overrun with inquilines (animals that live in the dwellings of other animals). While Kinsey could conduct some hybridization experiments, and could try to hatch wasps on different types of oak tree branches or rose bushes, gall wasps could not be produced as experimental, standardized animals the way that faster-breeding insects like *Drosophila* could. V. Betty Smocovitis's point that a scientist's choice of experimental animal "could therefore determine the kinds of evolutionary theories one supported or rejected" supports the idea that part of Kinsey's decision to leave the evolutionary synthesis was based on the inapplicability of gall wasps to the new forms of laboratory experimentation and evolutionary science inquiry.[55] Thus he took his scholarly inclination toward observation and synthesis to a field with study organisms more suited to his research processes and methods.

When Kinsey chose humans as his next experimental animal after gall wasps, he marked his academic career by choosing two kinds of creatures that were somewhat alike in their reproductive patterns. Neither gall wasps nor humans lend themselves to experimental breeding, and both kinds of animals are also slow breeders relative to other animals when they are reproducing, if they reproduce at all in a given year. Thus in the case of humans, Kinsey could separate sexual behavior and reproduction as two related but distinguishable processes of study. Of course, no scientist can be sure in advance that the organisms he or she has chosen will be suitable as a means of investigating all of their questions, especially their most general ones, and Kinsey could not foreshadow the development of the evolutionary synthesis that would effect a new development in his research. As "questions of style and personality enter into the decision whether to stick with an organism or switch to another in pursuing a particular problem," Kinsey saw that gall wasps could no longer answer contemporary evolutionary questions, and therefore he made the decision to change experimental animals to continue the field research in which he specialized, and to answer the questions about developmental processes that interested him the most.[56]

Though Kinsey did not pursue further research on gall wasps after the early 1940s, he stayed involved as the evolutionary synthesis developed a formal academic and disciplinary structure. The synthesis's disciplinary organization evolved at the same time that life science academic departments were also reorganizing themselves along the lines of research practices rather than objects. As Joe Cain has written, "Disciplinary transformations and shifting balances from objects to processes are familiar tropes in history of the life sciences across the nineteenth and twentieth centuries: out goes mammalogy, ornithology, herpetology . . . in comes physiology, ecology, genetics, and ethology."[57] In this period of multiple affiliations and shifting senses of priorities in the late 1930s, Kinsey certainly was not alone in using the opportunity to reassess his position in the academic sciences.

Kinsey maintained a broad interest in evolution by attending the first meetings of both the SSS and its postwar successor organization, the Society for the Study of Evolution (SSE). Kinsey's presence at the founding meetings of those organizations and his many colleagues in those groups kept him in contact with life scientists, their ideas and research interests, and tendencies toward synthesis in their work. Kinsey actively continued to maintain connections with other life scientists in his generation through his transition from one body of work to another. The movement from studying objects to studying processes in evolutionary theory also reorganized the disciplinary structures of academic biology. Consequently, in the late 1930s and early 1940s, life scientists became less interested in the life histories of individual specimens and more interested in the processes that created them. Kinsey's work in the human sciences reflects this trend, as his studies of human sexual behavior did not outline the sex lives of individuals, but rather described the physiological and cultural processes that structured the lives of different human groups.

In the spirit of turning scholarly attention to evolutionary processes, some life scientists, including Kinsey, Theodosius Dobzhansky, Simpson, Anderson, and Sewall Wright, met at the 1940 American Association for the Advancement of Science (AAAS) meeting to form a new organization for studying speciation. The group accomplished little during World War II, but an initial member survey showed members' enthusiasm for synthesis and for the integration of evolutionary studies. As the Society for the Study of Speciation's secretary, Alfred E. Emerson, wrote: "The need is felt by many students of speciation for a greater degree of integration between the various fields."[58]

Emerson collected the feedback of SSS members regarding the mission of the new society into a booklet soon after the founding meeting. It was full of quotations from members excited about the opportunities for collaboration in their approaches to evolution. The genetics professor J. A. Jenkins at the University of California, Berkeley, wrote that "the association might serve as a clearing-house for information and to bring together individuals and groups now scattered throughout the country." John Muirhead Macfarlane at the University of Pennsylvania seconded that desire, stating that the SSS should "secure close cooperation and common enthusiasm between workers in this country as well as abroad." Others suggested that the SSS coordinate local groups across the country, enhancing local, national, and even international cooperation. J. R. de la Torre-Bueno of the Brooklyn Entomological Society thought that Kinsey's material on gall wasps might serve as a base for bringing geneticists and taxonomists together.[59] Kinsey did not end up pursuing the synthesis of taxonomy and genetics in his own work, though he had numerous colleagues who were actively engaged in such projects. He thus began his sex research in an academic world that was moving toward collaboration instead of specialization, toward processes of evolution rather than discerning the objects those processes created.

The activities of the SSS were naturally limited during World War II, but once the war was over many of the same evolutionary scientists were eager to regenerate their collaborative spirit. Kinsey attended the first meeting of the reformed SSE in March 1946 at the AAAS meeting in St. Louis, Missouri, along with fifty-seven others. However, he was too far along in the work on his first compendium of human sex research, *Sexual Behavior in the Human Male*, to return full-time to the study of evolution.[60] The meeting of the SSE led shortly thereafter to the establishment of the journal *Evolution*, spearheaded by Ernst Mayr, to bring together taxonomists and geneticists through collaborative work on an interdisciplinary journal.

Kinsey also surfaced as a possible member of the Committee on Common

Problems of Genetics, Paleontology, and Systematics due to his membership on the executive committee of the SSS, but he did not become involved in it. So as Kinsey was beginning to assemble material for the *Male* volume, the atmosphere of scientists with different approaches to biology, working with each other toward common goals, was part of his inspiration to continue his own work of synthesis. Kinsey's primary contribution to the meeting was to advocate for a new society that would stand apart from the ASN.[61]

Kinsey stayed marginally involved in the newly formed academic structures of the modern synthesis through 1946, but work on the *Male* volume overwhelmed his ability to keep current with evolutionary science developments. His interests had developed in an alternate but related direction. His colleagues in laboratory-oriented evolutionary science, however, stayed in touch with him. Many of the most prominent contemporary evolutionary scientists, including Mayr, Julian Huxley, Anderson, and J. B. S. Haldane, gave Kinsey their sex histories to include in the *Male* volume. Kinsey's work so intrigued Haldane that he regularly contributed a sex diary to the Institute for Sex Research, arranged to meet Kinsey in London when the latter visited Europe in 1955, and wrote an obituary for him.[62] The evolutionary scientists in Kinsey's intellectual circle, who were interested in both life and human sciences themselves, recognized the blend of disciplines in Kinsey's new work and maintained an interest in it. Kinsey, for his part, moved from one interdisciplinary sphere, evolution, to another, human sex research.

Kinsey's adherence to the scientific worldview he had developed in his early career is evident not only in his textbook writing and in his decision to leave evolutionary studies, but also in his college-level teaching. Kinsey showed that he could skillfully synthesize the many facets of modern biology into his textbooks and workbooks, and that he could impart a fascination with and an interest in the natural world to high school students. His college-level science teaching, analyzed next to his marriage course teaching, shows more similarities than differences. Read together, evidence from his lecture notes on taxonomy, biology pedagogy, and evolution, and from those on the marriage course show that Kinsey brought those perspectives on life sciences to the human science of the marriage course. In both cases, he encouraged students to be explorers—to not "stay indoors" in their thinking. He encouraged them to trust their own observations, to weigh arguments by seeing all sides before making a judgment, reading widely, and comprehending, if not agreeing with, theories not based in physical fact. He wanted students to have a firm basis of understanding any theory or concept they were rejecting, and why. In short, he wanted them to become good academics, good scientists, and good citizens.

Kinsey's interest in studying the world as it appeared to him, along with teaching and learning from those around him, drew him into studying a new field where he could put his skills in collecting, observing, and categorizing data to best use: sex research. His focus on taxonomy, emphasis on instructing proper scientific method through textbooks to young people, his networking with mostly like-minded collaborators, and his insistence that elementary biology and some form of scientific method were both understandable and necessary for thriving in everyday adult life, carried directly over into his next educational project, the IU marriage course, and into the two volumes on human sexual behavior to follow. As Joe Cain puts it, Kinsey's biology textbooks, workbooks, and teacher instructional manuals all together "offer a wonderful illustration of 'synthesis' in every dimension."[63] And when the opportunity to teach, gather, and study in a new field appeared to him, he did not let preexisting disciplinary boundaries hold him back. He jumped right in.

3

Teaching Life and Human Sciences

I hope to prove to the world someday that any subject may be a profitable field for scientific research if zealously pursued and handled with objective scholarship.
—Alfred C. Kinsey to Ralph Voris, July 6, 1939

THE COMBINED FORCES of Kinsey's scientific writing paired with his biology teaching moved him into the world of human sex research during the 1930s. Together, his writing and teaching on life sciences soon led to his greater interest in exploring the human sciences. His teaching general biology at the college level led to his incorporating human biology and reproduction into his biology courses for Indiana University students in the early 1930s. That teaching decision, in turn, nurtured his interest in sex education at the postsecondary level, culminating in the first session of the marriage course in the summer of 1938. As his commitment to the marriage course deepened, he started asking course participants and then homosexual men and their heterosexual friends in Chicago and northern Indiana systematic questions about their sex lives. This chapter examines Kinsey's life science teaching in the 1920s and 1930s at IU, the ways that his life science teaching cultivated his research and preparation for the marriage course, the two-and-a-half-year history of the marriage course, and his initial forays into interviewing individuals outside of the IU community. The framework he built for asking questions and organizing data, along with the sex history interview form, became part of his method for mass collection of personal sex histories. This chapter shows how he brought together his interests and skills in science teaching, data gathering and ordering, and making connections with a wide variety of people to gain their confidence and support, in order to provide an intellectual structure for the intensive sex research that structured the next decade of his life.

The brief comment Kinsey made in his graduate school notebook, that the entomologist "should have a bigger side interest," would become even truer for him in early 1938 as he took the lead in organizing IU's marriage course. The marriage course proved to be an important moment in Kinsey's intellectual history. Fascinated by the intimate sex histories of course participants that he had begun to record, and discouraged by the shift to laboratory work in evolutionary biology, he turned more fully toward human sciences.[1] He gave up teaching the course after seven sessions in September 1940, in order to focus on the initial data for what would become his two largest and most comprehensive works, *Sexual Behavior in the Human Male* and *Sexual Behavior in the Human Female*.[2] To prepare those volumes, Kinsey needed to draw deeply on his abilities as a teacher and scholar of the life sciences.

Teaching Future Educators the Life Sciences

Forming future high school biology teachers was one of Kinsey's goals as a college professor. Kinsey taught biology and taxonomy at undergraduate and graduate levels from his arrival at IU in 1920 until the late 1930s. He then taught evolution to biology graduate students from 1935 until 1946, well into the preparation for the *Male* volume. Kinsey taught general biology, methods in biology for school teachers, taxonomy, and zoology in addition to teaching evolution. Kinsey's teaching and exam notes show that he aimed to give students a broad exposure to issues in biology, zoology, and evolution; that he kept pace with current developments in the sciences and taught them to his students; that he required students to participate actively in class; and that his courses included material from both the life and the human sciences. Kinsey's classes tended to be small, and as such facilitated hands-on intensive learning. "Seminar teaching," as Kinsey was doing, "was adapted by scientists to the needs of their own disciplines. . . . Students were actively inducted into the craft and standard of their specialties—in the laboratory, the botanical garden, the observatory, and the field, as well as in the seminar room."[3]

At IU, Kinsey offered either one or two classes per semester and one or two in the summer from 1920 through 1946. A course outline for a biology pedagogy class taking place from February to March 1930 shows that Kinsey taught future elementary science educators the differences between "pure" and applied science, how to develop teaching objectives, how to teach the scientific method, how to interest young students in science, and how to score exams. The class also included a field trip to a greenhouse.[4] In the school year 1940–41, Kinsey offered two summer session courses (a field course with twenty-seven students and an entomology lecture with ten students), an upper-level course on evolution with nine students in the fall, and two courses in the spring: taxonomy, with twenty-seven students, and an upper-level methods in biology education class, with ten students. In the fall 1941 evolution class, students signed up to present "special reports" to Kinsey and to their classmates on a wide-ranging list of topics, including geographic isolation; cytological isolation; the significance of mutation in evolution; paleontological data on evolution; the evolution of main plant groups, arthropods, or vertebrate classes; and human evolution. They also had to present book reports on classic and current evolution texts, such as Charles Darwin's *On the Origin of Species* (1859) and *The Descent of Man* (1871), Henry Fairfield Osborn's *The Origin and Evolution of Life* (1917), and Thomas Hunt Morgan's *The Scientific Basis of Evolution* (1932).[5] Kinsey encouraged students to do their own research and critical thinking on topics that he knew well and had long interested him.

One of the exam questions for a biology class designed for secondary school teachers indicates that Kinsey was concerned with instructing those future teachers to educate their own students about human sexual behavior as early as 1932. One of the questions Kinsey asked on a final exam, covering several of his primary interests, was "discuss the teaching problems presented by two of the following: classification, sex instruction, evolution, [and] choice of animal dissection material."[6] "Sex instruction" in this exam was simply another one of the topics that future biology teachers needed to cover, and Kinsey wanted to ensure that they treated it as evenhandedly as they did classification and dissection. He also wanted students to be as current on scientific literature and teaching methods as he was. The belief that young people learned best by observing and doing, rather than by reading or classroom lecturing, was safely ensconced in biology textbooks and teaching notes. It would take on a different resonance in the IU marriage course and then again in *Sexual Behavior in the Human Male* and *Sexual Behavior in the Human Female*. Themes and language that appear regularly in the *Male* and *Female* volumes appear first in Kinsey's teaching texts: the importance of experience to learning and trusting scientific facts and observation above social conventions.[7]

Seventy-five pages of teaching notes for his evolution class survive, along with exams. The notes, originally composed sometime in 1935, show Kinsey's engagement with classical Darwinism as well as with the most up-to-date literature on the subject. He orchestrated a classroom debate on the role of natural selection in evolution and informed students about recent papers by Sewall Wright, Theodosius Dobzhansky, Leslie C. Dunn, and others who were experimenting with the genetic means of evolution on *Drosophila*, salamanders, and many other creatures. The evolution teaching notes in particular reveal his awareness of the wide dimensions of his field and willingness to teach it when many American colleges and universities were not regularly offering courses on evolution.[8] In "the new mode of seminar instruction" that science instructors adopted in the first decades of the twentieth century, "students internalized and calibrated standards for seeing, judging, evaluating, and argument. These were the habits of mind and body that by the early twentieth century had been instilled and ingrained in a generation of scientists."[9] Kinsey's intellectual networks with other scholars of evolution thus manifested in the classroom as well as in his gall wasp books.

Kinsey's evolution class notes parallel his gall wasp evolution work in how they show his reading and engagement with well-known taxonomists. The primary topics he covered in his lectures were individual variation, hybridization, mutation, genetic rearrangement, Lamarckism, cytology, the nature of species, evolution, polyploidy, isolation, and—closest to his own interest—the nature of higher categories in species. The articles he cites most frequently are by Dobzhansky, Dunn, and Wright.[10] After quoting Wright in "Evolution in Mendelian Populations" as stating that theories of evolution are best derived from "the statistical situation on the species," Kinsey points out: "The appeal to actual conditions in nature seems to escape him." In his discussion of species, he notes that they are "imaginary—the individual is the only reality," and reiterates that point in his lectures on human evolution and individual variation. In offhand comments throughout his notes, he repeats his preference for large sample sizes, only studying specimens that are actively breeding in the present, and the reality of nature's conditions over those in the laboratory or generated by mathematical theory.[11] He continues his "treatment of questions in a broad, biological way"—which includes life and human science questions, just as Wheeler had.[12]

While he continued his teaching schedule into 1946, until the work on the *Male* volume became too overwhelming and he had to stop, teaching was constantly on his mind as he traveled the country giving lectures, making new contacts, and collecting sex histories. He drafted a final exam for an evolution class on December 17, 1943, on stationery from Hotel St. George in Brooklyn, New York. He drafted a midterm exam on stationery from the Deshler Wallick Hotel in Columbus, Ohio, in October 1946. Those planned exams contain questions to students that show that Kinsey was not only teaching them to memorize scientific facts; he was also teaching them skills in how to make their best judgments about scientific theories. The October 1946 exam asked students to "discuss values of [the] following sources of data on evolution: paleontology, genetics, taxonomy, comparative anatomy, [and] philosophy"; to analyze arguments for and against Lamarckism; and to answer the question "What is a species?" from several different theoretical viewpoints.[13] He drafted those particular exams long after the marriage course was over, and they reveal his ongoing commitment to teaching students how to weigh and judge data that authorities put before them, even when he himself was one of the authorities. His strong desire to teach science to high school and college students, to show them how to analyze scientific questions, and to instruct them in its use and importance in their everyday lives as citizens were all evident in his biology classroom teaching and in a different kind of teaching situation: the IU marriage course.

The Marriage Course

There was a short article on the front page of the *Bloomington Daily Telephone* on June 23, 1938, entitled "I.U. to Offer Course in 'Marriage.'" Unnamed Indiana University officials praised the proposed faculty-run, twelve-session, non-credit course. "Dependence on our civilization is largely a matter of preserving the family on a high level. . . . The course on marriage to be offered at Indiana University is intended to help family conditions."[14] Next to articles on the recent hot weather and the comings and goings of IU professors, notice of the course likely generated little interest at the time in the small, south-central Indiana town. However, the course, initially taught by eight instructors on the university faculty and staff, would have a stormy history ending with its lead faculty member resigning from it. The marriage course was a watershed moment in Kinsey's intellectual history; fascinated by the intimate sex histories of course participants that he had begun to record, he gave up teaching the course in order to focus on the initial data for what would become the Kinsey Reports.

Marriage courses appeared on two- and four-year American college campuses beginning in the late 1920s, amid dramatic cultural change in the lives of teenagers and college-aged young people. Secular educators attempted to teach Christian morals and values while realizing that modern technology and urbanization had irreversibly transformed marriage. By 1938—when Kinsey began the IU marriage course—marriage education in high schools and colleges was a small industry, with approximately 250 US colleges and universities hosting such courses. Marriage courses often combined lectures and small-group discussion, or lectures and individual conferences with instructors as regular features. The first version of the IU marriage course in the summer of 1938 followed the standard format of other marriage courses throughout the United States, with one important exception, and echoed many established themes of the burgeoning marriage course movement and the scientific field of sexology. The course included personal conferences with the instructors, also a common practice. But Kinsey's lectures for the course soon began to diverge from standard marriage instruction rhetoric, as he heard from students how much they appreciated learning about the uniqueness of individuals' sexual anatomy and desires, and as he began to do more intensive research in sex, with data from marriage course students and from other nonstudent groups. As Kinsey discovered the diversity of sexual behavior among undergraduate and graduate students, faculty wives, and heterosexuals and homosexuals in Chicago and northern Indiana, his lectures opened broad questions about sexuality that the marriage course, with its obvious focus on improving nuptial bonds, was not designed to address. As changes in the texts of Kinsey's lectures show, over time he focused less on how a healthy sex life enriches and stabilizes marriage and more on a broad range of human sexual experience. While the question of sex in marriage would remain one of the foci of Kinsey's analysis, it would no longer be the only one.

Changes in Kinsey's lectures during the marriage course show his increased interest in applying his version of the scientific method to sex research, moving beyond what other scientists had done so far. The importance of a biological approach to social problems, which Kinsey had first learned from his graduate adviser, William Morton Wheeler, became increasingly clear to him as he gathered the sex histories of students (and others) and read more widely in the newly developing field. Kinsey constructed human sexuality, for his students and himself, using his conception of scientific method. Based upon more than twenty years of biological research, he had learned and had taught that a large sample size and broad geographic distribution were necessary if a research project were to yield significant results. Those precepts, combined with faith in the scientific method (observation, recording, and interpretation of naturally occurring entities and events), would be solid foundations on which he could base valid conclusions. As Kinsey had consistently taught sex education methods to college students and heredity to high school students, he was well positioned to bring the techniques of scientific method to bear on a topic he had already examined from a biologist's perspective. The marriage course was a pivotal moment in Kinsey's life, as he turned from studying gall wasps toward studying human sexuality, with his full complement of taxonomic skills intact. As he wrote to former graduate student Ralph Voris in late 1939, "We will prove to these social scientists [psychologists, psychiatrists, and sociologists] that a biological background can help in interpreting social phenomena." After he gave up the marriage course in September 1940, Kinsey was able to devote most of his research time to gathering sex histories, to downplaying the counseling angle of the individual conference, to refining his interviewing technique, and to figuring out patterns in his data.[15]

In early 1938, an informal IU student group, including leaders of the Association of Women Students, Pan-Hellenic Council, Inter-Fraternity Council, and Blue Key, contacted Kinsey about chairing a course in the spring. There were many reasons why some among the six thousand–member IU student body had become interested in hosting a marriage course. First, some students had likely become aware of the growing number of marriage courses at other campuses through reading the *Indiana Daily Student* "Collegiana" column, which compiled news from campuses across the country. Second, college-aged men and women were acquiring venereal diseases (mostly gonorrhea and syphilis) in increasing numbers across the country in the late 1930s, and students may have wanted advice on how to prevent infection before marriage or on how to manage the effects of the diseases. In February 1938, with the support of campus physician J. E. P. Holland, the *Indiana Daily Student*'s editorial board successfully argued that free Wasserman tests should be available at the student health center.[16] Thus some of the more sobering aspects of marriage and sexual intercourse may have been especially on students' minds, coupled with a desire to avoid such problems. Third, students had grown increasingly less interested in the kinds of vague advice offered by religion-oriented marriage educators, including the longtime campus marriage instructor and IU School of Medicine professor Thurman B. Rice. Rice taught students that couples should not learn anything about sex before marriage but instead should learn through trial and error afterward. The information in Rice's marriage course never went beyond vague comparisons of human reproduction to amphibian and mammalian reproduction. IU students also regularly complained in the *IDS* about the dullness and uselessness of Hygiene 101, a required one-hour health course for campus freshmen. Fourth, Kinsey was already discussing sex education in his biology pedagogy classes for secondary education students, and those students would have been aware of his expertise in the subject. There was a ready on-campus audience for a course on marriage that offered some clear and practical advice on its realities, including sex, law, economics, pregnancy, and family dynamics.[17]

The IU Board of Trustees approved the students' request on June 9, 1938, and Kinsey began the process of selecting other instructors. Part of their work as a group involved becoming familiar with literature on marriage education, family life, reproductive biology, and sexuality. The new chair was no exception, although he already had ideas on how to teach the last topic. Kinsey wrote letters requesting information from other college marriage educators and read transcripts of published college marriage course talks. Kinsey recruited seven other IU faculty and staff members to lecture on their areas of specialty, and he and his colleagues read across marriage- and sex-related literature together. After working through the suggested readings, the instructors had also previewed each other's lectures, as had many of their spouses, in order to offer feedback and critique before presenting them to course participants. The marriage course was ready for its first student audience.[18]

The lead instructor had done significant reading in areas related to marriage and sexuality so that he could share them with the other instructors. He apparently did his own bibliographic research and found nine texts on his own.[19] While some of the ideas in those texts did not take deep root in the extant lectures, they nonetheless show Kinsey's thought process as he sorted through the vast existing literature on young people, marriage, and sexuality in the 1930s to find the material that he thought would be best for the course's lectures.

One of the authors with the longest paper trail in Kinsey's work, beginning with the marriage course, was Katharine Bement Davis, General Secretary of the Bureau of Social Hygiene in New York City in the early 1920s. She compiled the results of white, college-educated women's answers to questions about their sex lives in her best-known text, *Factors in the Sex Life of Twenty-Two Hundred Women* (1929). The book contains simple statistical analysis of data about the sexual behavior of both married and unmarried women, including their use of contraceptives, masturbation, homosexuality, frequency and periodicity of sexual desire, and marital happiness. While Kinsey would avoid such nonspecific terms such as "happy" and "unhappy" as accurate measurements for people's sex lives, Davis's large sample size and thorough questionnaire nonetheless provided solid proof of several points critical to the themes of the IU marriage course. Davis found that premarital sex instruction greatly improved the chances of women's overall happiness after marriage; the women who knew more about sex before marriage were more likely to be sexually satisfied after their nuptials than those who knew nothing; and when women had positive sexual experiences on their wedding nights and honeymoons, they were more likely to be continually satisfied in their sex lives as they grew older.[20] Davis's work showed "natural sexual variation without invoking the psychologist's habitual urge to abnormalize it."[21] Kinsey would adopt many of Davis's conclusions about women's sexual behavior. He too came to believe that sexual instruction should be coeducational, that premarital restrictions on sexual knowledge were especially damaging to women, and, more generally, that the more people knew about sex before they engaged in it, the less marital discord they would experience.

William S. Taylor was another author whose work would also make appearances in the marriage course and Kinsey's future books. Taylor's "Critique of Sublimation in Males" was a direct attack upon the psychoanalytic concept of sublimation. Taylor defined sublimation as "a diversion of 'sex energy' into 'higher forms' of activity"; he found that "it may have had good results in encouraging idealistic efforts; but it has bad effects upon many single individuals."[22] In his face-to-face interviews with forty healthy, single, white male graduate students (followed by a written questionnaire several months later), he found that all struggled to be chaste, but that none were able to channel sexual desires into intellectual, spiritual, or athletic avenues. Only sex or sex play— masturbation, petting (heterosexual sex play without penile penetration), intercourse with women (prostitutes or not), and/or nocturnal emissions—would satisfy them.[23] In short, denying sexual desire was far more damaging to those men's psyches than indulging it, regardless of what social hygienists and moral reformers said. Taylor's primary findings—that a key psychoanalytic concept like sublimation did not stand up to empirical investigation, and that men were needlessly suffering as a result of denying their sexualND yearnings—would have a powerful impact on Kinsey's own analytical framework. Taylor's overarching message inspired Kinsey's lectures: "Prejudices, fears, wishes, pretenses, rationalizations, [and] amnesias, have allowed too little scientific knowledge of sex."[24] In the opening lecture of the first marriage course, "Biologic Bases of Society," Kinsey would criticize the prudishness of those who would hide truths of sex from the young. It was scientific knowledge, not prejudices, wishes, or fears, that Kinsey would impart to the marriage course students, and tipping one of the sacred cows of psychoanalytic theory was one aspect of his message.

Kinsey suggested that students read the books that had played a role in shaping his own thinking. For example, at the top of his lecture on "Reproductive Anatomy and Physiology," Kinsey lists two titles that he recommended to students, Bronislaw Malinowski's *The Sexual Lives of Savages in North-Western Melanesia* (1929) and Abraham Stone and Hannah M. Stone's *A Marriage Manual: A Practical Guide-Book to Sex and Marriage* (1935).[25] While it is clear from their comments that not many students actually read these two books, *A Marriage Manual*, authored by a married couple (both doctors) active in the New York City–based birth control movement, was by far the book that most students filling out their questionnaires named having read by the end of the course (38 out of 215 in the fall of 1938).[26] *A Marriage Manual*, reprinted and updated numerous times until the late 1960s, is a series of questions and answers between the authors and a hypothetical married couple. In language that Kinsey would later utilize, the Stones state that "an intelligent [marital] union should be based on an understanding of the biological processes involved," and they proceed to advise young people about sex and marriage in a straightforward style.[27] Their tone is just as important as the topics the Stones covered. They instruct readers in reproductive physiology; conception; birth control; the importance of regular, mutually satisfying sex to a healthy marriage; impotence; and sex during pregnancy or menstruation; among other topics, some of which Kinsey and his fellow marriage course instructor Robert L. Kroc would touch on in their lectures. While the Stones differed from Kinsey in some important respects—they maintained that sex should only occur within marriage—they advocated for the free dissemination of sexual knowledge without secrecy, embarrassment, or moral gloss.

Malinowski's approach to sex in *Sexual Lives of Savages* also influenced Kinsey's own. Malinowski, one of the few anthropologists in the 1920s who addressed sex directly with his subjects, focused on several themes that became key to Kinsey's thinking: that children should learn about sex among themselves with little interference from adults; that most couples had premarital sex and practiced a form of companionate marriage before officially marrying; and that orgasm for both the man and woman marked most successful heterosexual liaisons. "In the ordinary course of events," Malinowski wrote, "every marriage is preceded by a more or less protracted period of sexual life in common. . . . It serves also as a test of the strength of their attachment and extent of their mutual compatibility."[28] For Malinowski, sex before marriage improved couples' postmarital relationships. Further, he found that couples had intercourse in a variety of positions in a single session of lovemaking to ensure that both had orgasms. He notes that the man often held his orgasm back so that the couple could orgasm at the same time. Kinsey's marriage course lectures would consider each of those observations as true for his largely Midwestern, collegiate, and white audience.

Gerrit S. Miller Jr.'s article "The Primate Basis of Human Sexual Behavior" (1931) and Howard M. Parshley's *Science of Human Reproduction* (1933) resonated only lightly in the marriage course lectures but would more strongly influence both the *Male* and *Female* volumes with their anthropomorphic comparisons of primate and human sexuality. Miller, anticipating the animal behaviorist Frank A. Beach's work by a decade, argues that human sexual behavior is much like mammalian sexual behavior, particularly the behavior of primates such as orangutans, chimpanzees, baboons, and "several Old World monkeys." For humans and primates, "physiological rhythms are no longer the nearly exclusive regulating factors in the mating behavior of either sex. . . . [Mating behavior] tends, in both sexes, to assume more nearly the form of an ever-available amusement activity than that of a periodic blind submission to an inescapable racial force." Citing Gilbert V. Hamilton's study of macaques, among others, Miller states that "individual variations of the sexual behavior" are most common among this group of mammals, whereas others mate by rote instinct to propagate the species. Parshley was a friend of Kinsey's from graduate school at Harvard, whose work echoed sentiments similar to Miller. Kinsey signed his personal copy of Parshley's book, which states bluntly, "If we wish to understand human sex behavior, then, we must first turn to the animal world," with the date "1940."[29]

Miller uses mammalian data from other researchers to suggest that male and female masturbation and same-sex erotic play were normal for humans and primates. However, he believed that human men's desires both to have a steady mate and to stray from her occasionally, along with the possibility that men could force women to have sex without the latter's consent, set them apart from their primate cousins. Parshley similarly finds, in a passage Kinsey underlined three times, that "the life-long association of one male with one female is among the rarest of biological prodigies."[30] Miller reasoned that unlike all other species, human men rather than human women had control over sexual encounters because men had greater power in human society. Such power was due to the "overthrow," an idea put forth by the American sociologist Lester Frank Ward, among others, that men had gained control over women in "gynæcocentric" preliterate human societies by committing mass rape.[31] Kinsey did not ascribe to the belief in a gendered power differential based on unproven and ancient history, as Miller had. However, Miller's article exposed Kinsey to two key problems in any attempt to research the whole of human sexual behavior: how to account for differences in power and control over sexual behavior between men and women, and how to study rape. He would attempt to address both topics, especially in the *Female* volume, but would leave rape and sexual violence almost completely out of his narratives.[32]

A final important source for Kinsey, and one that would receive high praise from students—though he never mentioned the source in the lecture transcripts or put it on the faculty reading list—were the illustrations first printed in the obstetrician/gynecologist Robert Latou Dickinson's *Human Sex Anatomy* (1933).[33] When Kinsey first wrote to the obstetrician/gynecologist, apparently in praise of *coitus reservatus* (withdrawal) as a contraceptive method, he stated that Dickinson's work was one of his inspirations for turning to sex research from gall wasps.[34] However, they did not establish a working relationship until 1941, when Kinsey published his first sex-related article. When Dickinson wrote to Kinsey to praise the article, Kinsey replied, "You deserve unreserved credit for pioneering in this field, and consequently I am particularly encouraged by your favorable reaction to our first publication." He repeatedly thanked Dickinson for his expertise, willingness to share his own research, providing Kinsey with contacts, and for giving his sex history to the project. He also repeatedly claimed later that Dickinson was an inspiration to him.[35]

Human Sex Anatomy was intended for medical professionals, and it contained detailed textual descriptions and visual representations of male and female sexual organs, in both pen-and-ink and watercolor (though reproduced in black and white). Kinsey made lantern slides (glass slides that could be projected) from these drawings to show students during his lectures.[36] There were also sagittal outlines of different intercourse positions, focused on the genitals, showing the various angles at which the penis would enter the vagina for each. Furthermore, while it is unclear whether Kinsey read the text of *Human Sex Anatomy* before the marriage course, or just made slides from the drawings, the language and tone are similar to what he would adopt later in letters, lectures, and the *Male* and *Female* volumes. Dickinson wrote in the conclusion: "When the time comes that diagrams and descriptions of the normal functions of these universal organs are as generally acceptable and as much taken for granted as are pictures of other life processes, we shall doubtless wonder at our artificially fostered mystifications and many of our elaborately manufactured attitudes of shame, and yet not lose our feeling that privacy for intimacy is good taste as well as enhanced delight."[37]

Robert Latou Dickinson also encourages readers to join him in conducting research on "average sex experience" and "normal sex life": "It is time we begin building on detailed case records running through lifetimes in series counted in tens of thousands."[38] Kinsey took this suggestion to heart, as Dickinson and his books would play a notable role in Kinsey's research.[39] Dickinson, Malinowski, Taylor, and the Stones all thought that one day information about sex and reproductive anatomy would be more freely available. Kinsey shared that belief, as he railed against "prudish ideas" about sex and contraception in all three versions of the opening lecture of the marriage course.[40]

In summary, the IU marriage course faculty and student reading lists contained some articles and books Kinsey would later ignore and some that would become resources for the rest of his career. *Factors in the Sex Life of Twenty-Two Hundred Women* and *A Marriage Manual* would remain important to him, while others, such as *The Mothers* (1931) and *The Art of Being a Woman* (1932) barely factored in the philosophy he was developing about human sexuality.[41] From an analysis of Kinsey's reading list, it is clear that he was searching widely across marriage education and sex-related topics for literature that satisfied his requirements for scientific validity. He preferred that his fellow lecturers and students be most familiar with works that used face-to-face or survey interview data to make conclusions; had a relatively large sample size; treated sexuality with a minimum of moral, religious, and ethical framing; used precise and scientific language; and did not shy away from nonmarital and solitary types of sex. He rejected or used with caution works that were primarily the opinion of the author, made qualitative judgments about certain types of behavior or the persons engaging in it, or depended on psychological theories for their conclusions. He read texts that analyzed men and women as separate groups, as well as works that examined them interacting with each other. For the marriage course, Kinsey would mostly focus on the interactive data that he had gleaned from his readings, as that was the most relevant kind of data for the engaged and newly married just beginning to learn to live with one another.

The first notice for the summer 1938 course in the classified section of the *Indiana Daily Student* states its basics: "A non-credit series of twelve lectures on legal, economic, sociologic, psychologic, and biological aspects of marriage will be available for the first time during the Summer Session." The lectures would take place on Tuesdays and Thursdays at 7 p.m., and applicants (faculty wives, graduate students, seniors, and engaged or married younger students) needed to meet with Kinsey before the course began to gain admission. The location of the first session of the course is not clear, but from the fall 1938 session onward the sessions took place in the Chemistry Building auditorium.[42]

"Biologic Bases of Society," Kinsey's opening lecture in the first session of the IU marriage course on June 28, 1938, echoed the language he had used in *Methods in Biology* a year before to discuss teaching sex education in secondary schools. Kinsey began by pointing out to the course's ninety-eight participants that, in contrast to the significant amount of research on the reproductive behaviors of social insects, research on human pairing and sexual behavior was comparatively scanty, and that general information about it was more often based on "gossip and guesses" than on scientific fact. Marriage, he went on, was necessary to protect and to raise children, but delaying sexual intercourse often psychologically damaged those who remained virgins until wed. Delaying marriage until the mid-twenties, as was increasingly common, not only precluded adjusting to another person's sexual desires and becoming familiar with one's own, it also made adjusting to married life in general difficult. Given that taboos against premarital intercourse were unlikely to change any time soon, and that "adequate and mutually satisfying means of contraception" were unreliable or hard to obtain, the marriage course would provide tools for students to tackle their difficulties. Kinsey concluded his lecture by telling students, "Each man in his own field will present something of the special material which will provide the material by which you can work out your own solution."[43] While he did not explicitly advise students to experiment with sex before marriage, many students heard this lecture as a tacit admission that doing so would be good for their health.

Kinsey's second lecture, "Reproductive Anatomy and Physiology," was the fifth in the series. He showed black-and-white lantern slides of male and female embryonic genitalia in development, then genitals of mature men and women. He informed students that coitus was impossible without erection, but that women did not need to be aroused for intercourse to occur; and that the clitoris, not the vagina, was the primary source of women's stimulation, although many men were unaware of that. The next slides were of penile erection, aroused male and female genitals, and different coital positions, all from Dickinson's *Human Sex Anatomy*. Kinsey briefly covered pregnancy prophylaxis (tubal ligation, vasectomy, condoms, and diaphragms) but warned his audience that his lecture was only a glimpse of what sex was and what it meant in marriage: "It is quite possible to know all that need be known about the anatomy and physiology of reproduction and still grasp nothing of its art, but our excuse for bringing you this much of such material is a conviction that absolute ignorance makes it impossible to become a master of anything. So I give you this much and we [the other instructors] give you the material in the later hours with the conviction that knowledge can do no harm and may be the means of working out adjustments that are fundamental."[44]

At the end of this lecture, Kinsey reiterated the theme of his first address,

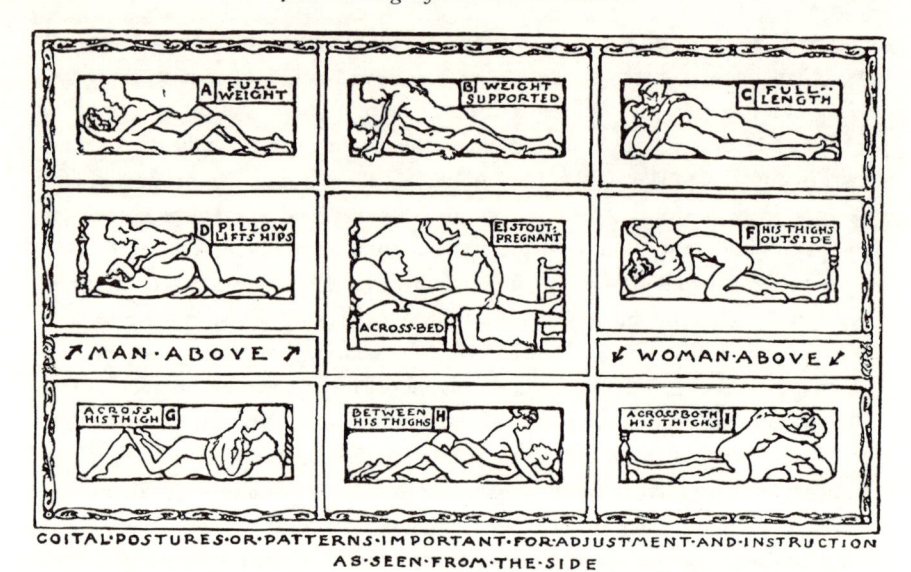

COITAL·POSTURES·OR·PATTERNS·IMPORTANT·FOR·ADJUSTMENT·AND·INSTRUCTION
AS·SEEN·FROM·THE·SIDE

Figure 3.1. Drawing of coital positions shown in the Indiana University marriage course. Plate 159c in Robert Latou Dickinson, *Human Sex Anatomy: A Topographical Hand Atlas*, 2nd ed. (Baltimore: Williams and Wilkins, 1949). Reproduced by permission of Wolters Kluwer Health/Lippincott Williams & Wilkins.

that the "art" of sex was essential to happy marriages. Ideas on how to learn that art might be found in books such as *A Marriage Manual* and *Human Sex Anatomy*, but Kinsey subtly placed more emphasis on "working out adjustments" in practice than on reading about them. "Reproductive Anatomy" was as explicit as lectures on sex in marriage education courses got, mirroring lectures by physicians such as Raymond Squier at Vassar College. In this lecture, however, Kinsey left the question of when students should actually learn the art of reproduction—before or after marriage—unanswered.

The zoology professor Robert L. Kroc gave the sixth lecture of the course, "Endocrine Basis of Sex and Reproduction." With accompanying slides, Kroc covered men's and women's disparate hormonal development, menstruation, ovulation, the mechanics of how women become pregnant, and menopause. He also dismissed the concept of the "safe period," a time in a woman's menstrual cycle during which she could supposedly have intercourse without risking pregnancy. Condoms, diaphragms, or abstinence were much more reliable choices for preventing pregnancy. "If you consider the fact that spermatozoa may live in the female reproductive tract for as long as two days . . . the so-called 'safe period' or rhythm idea for the control of conception seems to me at least to be hardly practicable because of the uncertainty of predicting for any particular person the time of ovulation," he stated. In a foreshadowing of Kinsey's "Individual Variation" lecture, Kroc addressed how the recent (1934) discovery of "male" hormones in women and vice versa indicated that the sexes were not as disparate as scientists had previously thought: "I hope that you have understood [from] this and from Dr. Kinsey's lecture that sex is actually a blend. Each individual, male or female, represents a blend of male and female characteristics."[45] Together, Kinsey and Kroc were building a portrait of a marital world in which a healthy sex life was foundational, mutual orgasm was a consistent goal, both partners understood the functioning of each other's bodies, and both understood how to encourage and to prevent pregnancies. Both lecturers emphasized that couples could work together to solve marital and sexual problems. For young men and women who might have worried about their normalcy compared to their peers, Kroc's lecture would have provided some reassurance that they were not alone in their desires, feelings, and behaviors, and Kinsey's last lecture would do so as well.

In "Individual Variation," the seventh lecture in the series, Kinsey emphasized the many variations of human genitals and sexual behavior. On the classroom chalkboard, Kinsey drew graphs of average clitoris and penis lengths, showing how in some cases their lengths overlapped. His aim in doing so was to show how men and women diverge both within and between the sexes, and to demonstrate that those divergences blur rather than reify sexual differences. Furthermore, Kinsey advised, there was no such thing as abnormal behavior, as "nearly all of the so-called sexual perversions fall within the range of biologic normality." As many men were ignorant of the fact that female orgasm was comparable to male orgasm, they needed to learn that most women needed manual or oral stimulation of the clitoris to ensure sufficient lubrication for penetration and female orgasm. While women could have intercourse without being aroused, as Kinsey had stated in the "Reproductive Anatomy and Physiology" lecture, if their husbands were skilled lovers there was no reason that they should have to. Knowledge of individual variation in marriage would lead to mutual respect and understanding as partners worked patiently toward achieving what Kinsey posited as a significant goal of marital coitus—not children, but simultaneous orgasm. This lecture, like "Reproductive Anatomy and Physiology," informed students that their bodies and thoughts were normal, that it would take time to reach a mutually satisfying pattern of behavior, and that orgasmic equality (with the implied use of birth control) was a crucial element in successful marriages.[46]

After the second-to-last lecture by the psychologist Harvey Locke, Kinsey passed out questionnaires to the students. The questionnaire, Kinsey told the students, would allow them to provide their "reactions on each and every point of the course, suggestions for handling the course in the future, the technique of the administration of the course, etc."[47] He compiled the anonymous (except for gender) data from all ninety-eight, recording his aggregate results on a copy of one of the forms. Kinsey's report to IU's new president, Herman B Wells, contains the first evidence that he was expanding his investigations of human sexual behavior beyond research for the marriage course. Kinsey notes that he had met with thirty-two students individually regarding "marital problems and personal sex adjustments." Kinsey was particularly excited about the case histories that he had collected from those meetings. He wrote to Wells, "You will be interested to know that the personal case history work bids fair to become one of the most significant parts of the program. The 32 cases handled by the biologists this summer was a startling indication of the need of such work on the campus."[48] His passion for gathering original data was quickly shifting from gall wasps to sexual histories of marriage course participants, which he solicited at the end of lectures and conducted in anonymous privacy. The histories became the core of his sexual behavior data after his extensive reading. According to Pomeroy, "these early interviews covered age at first premarital intercourse, the frequency of this activity, and the number of partners," among other basic facts of a person's demographic history, such as their age and family history.[49]

The idea of private conferences did not originate with Kinsey, but he modified the practice to suit his data-collecting propensities. The most prominent figures in marriage education—including Ernest Groves (University of North Carolina, Chapel Hill), Joseph Kirk Folsom (Vassar College), and Henry Bowman (Stephens College)—used the personal conference as part of standard marriage course practice. Kinsey differed from other marriage education instructors not because he conducted personal conferences during his marriage course, but because those personal interviews came to take priority as a means of gathering scientific data on sexuality.[50] Further, while other marriage educators used the private conference to give students advice, Kinsey did not—he asked only what they did and passed no judgment.

Kinsey kept his co-lecturers and the university president apprised of the progress of the private interviews after each section of the course ended. Before the fall series began, Kinsey shared the most recent student evaluations with the marriage course staff. He also updated his friend and former graduate student Ralph Voris on the wealth of information he was finding in the individual case histories. Kinsey's records and correspondence make clear that, while some of the sixty-four men and thirty-three women brought more than one problem to the private meetings, a handful of issues preoccupied most of them. Fifty-nine of the students, he wrote, wanted advice on petting, twenty-nine on premarital intercourse, thirty on reproductive anatomy and physiology, and thirty-six on masturbation. Nine also asked about homosexuality. As student and public interest in the course grew, the histories were beginning to dominate Kinsey's attention.[51] Enrollment in the fall of 1938 more than doubled from the summer session, to 217 students. That session had 136 seniors, 32 graduate students, 10 juniors, 4 freshmen, and 16 nonstudents, such as faculty wives and local townsfolk. Ninety-seven participants were women and 108 were men. A growing number of students (seventy-eight) had availed themselves of the opportunity of an individual conference with Kinsey by the end of the lecture series.[52]

Kinsey was a busy man in the summer of 1939, as he was not only leading a fourth session of the marriage course but also beginning to travel to Chicago and northern Indiana to start another sort of research—gathering the sexual histories of homosexual men and their heterosexual friends.[53] Inspired by the sex histories he had already taken in the marriage course, he set off to find more, and a major city with a large military and transitory population was a logical place to start. His correspondence files contain letters to and from about twenty men and women whom he interviewed, and they in turn provided him with new contacts among their friends and news about the underground homosexual community in Chicago. He befriended a married couple who operated a boarding house at 711 North Rush Street and took their histories and those of their tenants. One male correspondent informed Kinsey that there was a raid on "Club Gai" in the summer of 1940, in which 150 men and women were arrested fifteen minutes before closing time. The raid had a chilling effect on queer social life in the city; according to the letter-writer, "Guess innumerable girls and boys have 'taken the veil' for the time being. Poor kids." Kinsey, for his part, reassured the man: "Above all, you must understand that I, personally, harbor no disapproval of anything that you or any of the other folk have ever done or will do."[54]

Kinsey's correspondence with his contacts in Chicago shows that he was starting to think through problems of sexuality that would occupy his attention for the coming years. Whether nature or culture had a stronger effect on human sexual behavior was primary among them. One man, about to enter the military, wondered if he would be able to train or to force himself into being heterosexual so that he would not be subject to a dishonorable discharge. Kinsey responded that more people would probably have sex with both men and women if society did not shoehorn them into a choice of only homosexual or heterosexual identities: "Social factors do a great deal to force an individual into an exclusively heterosexual or homosexual pattern. Most of the social forces encourage the heterosexual, but society's ostracism of the homosexual forces him into the exclusive company of other homosexuals and into an exclusively homosexual pattern. Without such social forces, I think most people would carry on both heterosexual and homosexual activities coincidentally."[55]

With the evidence from his Chicago contacts and the IU students at the forefront of his mind during the marriage course, Kinsey leaned more toward culture affecting sexual behavior than a person's physical or physiological makeup.

While he considered the job of educating students about marriage important for their health and well-being (in one instance, Kinsey requested that an IU physician provide ongoing treatment for a student with gonorrhea), the data that he was gathering about human behavior from them and from others like the Chicagoans—and his method for collecting it—had the potential to make a broader impact on American society. He had come to an understanding of the potential impact of his own work, and he was beginning to envision how an objective, scientific approach could transform the study of sexuality beyond marriage and take it to a wider audience. Kinsey wrote to Voris:

> Then, of course, this marriage course program has prospered and multiplied work. In the first four semesters we have had 100, 200, 230, 260=790 students. A few flurries with unfavorable criticisms from older faculty who had no firsthand knowledge—but even that is gone. The students would do anything for us, their appreciation is so great. We have their written comments at the end of each semester. Several have written personal letters to express their appreciation for their personal benefit. Following your suggestion, we have tapped fraternity house gossip and find the course treated *most* considerately. The Gridiron [football] banquet brought only one reference to it—a reprimand to a couple of boys for having engaged in biologic activities "without benefit of Kinsey's course in connubial calisthenics." . . . [The course] has given us a wealth of material by which, Mr. Man—I hope to prove to the world someday that any subject may be a profitable field for scientific research if zealously pursued and handled with objective scholarship.

While Kinsey was willing to pass on a joke about the course to Voris, he also saw the potential to add to scientific knowledge about sexuality, "if zealously pursued and handled with objective scholarship."[56]

The knowledge that Kinsey was garnering through collecting more diverse sex histories, along with the trust he gained from the students he had interviewed, began to affect how new students in the course perceived its meaning to them. The fall 1939 course began Monday, September 23. Each *Indiana Daily Student* article about the course mentioned that the marriage course staff was available for personal conferences, that students were able to anonymously evaluate the course, and that very few of the 350 or so marriage courses across the country had as many different disciplinary perspectives as IU's course did.[57] The student responses to the fall 1939 lecture series as a whole were highly positive. One man was especially glad to hear "Reproductive Anatomy and Physiology," as "This is the lecture most of the students look forward to and for many it is the most valuable. I have heard many students commend your objective attitude." The sex instruction lecture struck many students forcefully, and they realized how poorly it would play to an audience of their elders. A man asked, "Why not give this lecture to incoming freshmen? Or are the University officials too prudish to be honest and admit the need of this information?" Another thought, "Most interesting and would make many of the 'old school' very angry, including many parents." A third posed several of the questions that would be at the heart of *Sexual Behavior in the Human Male* and *Sexual Behavior in the Human Female*: "Can you not give more accurate information concerning the need for outlet? If the need varies with individuals, how can one judge how great the need is? What biological effects result from too little or too great usage of outlets?"[58] By contemplating these broader questions, students grasped, perhaps along with Kinsey and the other instructors themselves, the potential impact that the information given in the course could have not only on the IU campus but also on the lives of numerous others. Student participants had begun to broaden their perspectives about possible ramifications of the course. If someone could confidently answer questions about how individuals were actually managing their sexual desires, perhaps the last student quoted above would better be able to decide how much "outlet" he should have. By having statistical data at hand, Kinsey's statement that "you can work out your own solution" would not make the moral decision to have sex or not any easier, but such data would, at the very least, show how other people had worked out their own solutions.

As outside criticism of the course grew, the students' defensive posture regarding the course would only become stronger, as exemplified in the comments for the spring 1940 session. Change was definitely in the air, as the sociology professors (including John H. Mueller, a Kinsey foe and the husband of the dean of women Kate Mueller) had resigned from the marriage course as a department. They complain in a letter to President Wells, "The importance of the work of the sociologists in the course is much less than had been anticipated." Despite that ominous prelude, the spring 1940 session proceeded, with Kinsey giving revised versions of the four lectures that he had given in previous sessions (he added a fourth, "Sex Education," in the fall 1939 session). In "Bases of Society," he asserted that "individuals can reach their finest development as a result of marriage." Such optimism about the place of marriage in the human life cycle had been less prominent earlier in the course, suggesting that Kinsey took seriously students' beliefs that marriage could indeed be a means to happiness. However, "Reproductive Anatomy and Physiology" revealed distinct changes in Kinsey's thinking and teaching over a two-year period. As he had stated concerning the basics of coitus in the first summer 1938 lecture: "It is quite possible to know all that need be known about the anatomy and physiology of reproduction and still grasp nothing of its art."[59] At that time, Kinsey had not gone into detail about how to learn such an art, beyond recommending *A Marriage Manual* and hinting that premarital experimentation was wise. In the spring 1940 session, Kinsey spent more time on the mental stimulation accompanying arousal, foreplay, and intercourse—not precisely a discussion of the "art" of sex but of how attraction begins and then changes over time: "Erotic response may be brought about by psychological stimuli. You may get this from a mental situation which has no physical contact involved. In the human erotic reaction, we depend on a combination of physical and psychological stimulation. The capacity of an individual to respond to a psychological stimulus depends on the previous experience the individual has had, [and] the set that he has toward it, so that ultimately there may be built up such mental associations around sexual contacts that the psychological stimulus alone will bring forth erotic response."[60] In Kinsey's formulation in this lecture, sex starts in the mind, proceeds to the rest of the body, but then returns to the mind to be part of the next experience. The more positive experience a person has or even desires, the more chances he or she will be mentally ready for the subsequent encounter. Conversely, if a person has had poor sexual experiences, those would influence their behavior as well. In the space of a year and a half, Kinsey had shifted from suggesting that premarital sex would make marital sex more pleasurable—and blaming "prudish ideas" for sexual unreadiness—to testing broader theories on the development of human arousal and sexuality in general. Kinsey's findings on the psychology of sexual response would take an entire chapter of *Sexual Behavior in the Human Female*, and the chapter's introduction would use nearly identical language.[61]

The same lecture also included, for the first and only time in the marriage course, Kinsey's thoughts on the relationship between sex and love. Even as he impressed upon students the importance of love and happiness in inspiring and sustaining marital relationships, he insisted that feelings and emotions could not withstand the scientific scrutiny that sex could. Whatever his personal feelings on the subject, he had little use for love as a point of scientific inquiry: "Now intercourse consists of a series of physiological reactions which are as mechanical as the blinking of an eyelid. . . . Emotional acceptance of this series of physiological events may be what you recognize as love. It may provide the inspiration for the writing of the poet, and the philosopher, of all mankind, but fundamentally, at base, the first part of the story is a story of mechanical responses which are as inevitable when the stimuli are provided as any other ordinary reflex of any other part of the body." In this analysis, love is, at its most fundamental level, an emotion that people attribute to a natural desire for intercourse. Perhaps attaching feelings to sex is unique to humans, but the desire for intercourse and reproduction is not, as Kinsey illustrated through the example of male moths who follow the scent of females from miles away in order to mate with them.

Love was not physiological, but it could become part of the "mental associations around sexual contacts" that people would bring to their sex lives. Nor did love guarantee or even portend a mutually satisfying sex life in marriage. Finally, Kinsey added that while birth control may have separated sex from procreation, it did not separate idealizations of love from marriage. In this version of the "Reproductive Anatomy and Physiology" lecture, Kinsey began to articulate one of the guiding principles of his future research: to understand the nature of human sex behavior, it was necessary to divorce the academic study of sexual behavior from love. Behavior and even arousal were quantifiable phenomena that could be observed and studied objectively, but love was not.[62]

The 1940 version of the lecture on "Individual Variation," though framed in a marital context, was not much concerned with the wedded life. It is a thumbnail sketch of how Kinsey was beginning to tackle problems beyond the scope of the marriage course. Kinsey articulated three additional guiding principles of his sex research: first, that the only kinds of abnormal sex were none at all or those which caused harm; second, that all other forms of sex, however rare, were simply variants on the complex continuum of human behavior; and third, that sex researchers were likely to bias their results according to their own values, and thus should make a special effort to be neutral:

> There is practically nothing in human sexual behavior . . . which deserves the term abnormal in the sense that it interferes with physiological well-being. There are cases when you might label a phenomenon abnormal in the sense that it interferes with the well-being of the species and in connection with that the only sexual abnormalities are celibacy, refusal to marry; abstinance [sic], failure to have intercourse; and delayed marriage, and that is a very different list than is ordinarily given in the books. In actuality, the classification into normal and abnormal that is usually made merely represents the type of behavior that the classifier has not happened to engage in.

Further, Kinsey identified the problem of understanding the interrelationship of the sexual body and social culture, stating that the reasons behind human behavior are never easily explainable: "It is one of the most difficult factors for the student of biology . . . even more so for the student of sociology to determine how much is environmental and how much is acquired." Even a dozen years later, when he published *Sexual Behavior in the Human Female*, Kinsey felt he still had not solved that fundamental problem to his satisfaction, but neither had anyone else.[63]

After painting sexual variation with the broadest brush possible, including hinting at the normalcy of taboo behaviors such as homosexuality and masturbation, Kinsey spent the remainder of the "Individual Variation" lecture on adjusting to sex in marriage. The latter part of this lecture also reveals his thinking at the time about the different roles of men and women in a heterosexual couple's sexual life. He repeated his belief from previous lectures on this topic that simultaneous orgasm was "more of a factor in holding the home together than is often realized," as it provided the ultimate satisfaction for both spouses and a way for them to focus wholly on each other. Achieving a happy and harmonious sexual adjustment between two differently desiring individuals, or what Kinsey would later call in the *Male* volume "sufficiently frequent and emotionally effective intercourse," "represents a compromise rather than a matter of capacity." He was discovering that the women he interviewed tended to want less sex than men, but it was not clear if that was due to heredity, environment, leftover Victorianism in American culture, men not even realizing their girlfriends or wives could orgasm (much less helping them do so), or some combination of those factors. Each couple would have to "work out [their] own solution."[64] The "Individual Variation" lecture ended with Kinsey quoting a couple who had come to him for sexual advice and had kept in touch. Kinsey reiterated his main points through them: that sexual adjustments take time, effort, and practice, and that couples can adjust to each other's desires if they are willing to do so. Thus individual conferences and sex histories clearly had had a significant effect on Kinsey's methods for teaching sexuality to IU students, as he directly referenced them in this lecture as a source for his new findings and new questions.

Kinsey's last lecture in the spring 1940 series was "Sex Education." He was aware that many in the student audience had never heard a forthright lecture on the subject, and so he designed the talk as much for the young adults as for their future children. He began by arguing that sex education should be given to children by parents and not "experts." Parents should begin sex instruction when the child was between five and seven years of age, when sex had no erotic connotation and before he or she would begin to learn about it from other children. Both boys and girls needed to know that having sexual feelings was a healthy sign of growing up, and that sublimation of such feelings usually caused emotional harm: "I have seen people who are completely unstrung by the attempt to avoid it and get along without sexual outlets," Kinsey declared. Boys especially needed to learn that masturbation was not harmful, as they had on average 3.5 orgasms per week in adolescence, compared to less than one for girls. "If you can set your boy straight on that subject you will have saved them eight years of worry," Kinsey assured his listeners.[65]

After covering the basic problems faced by adolescents, Kinsey described to the students how he had begun to collect sex histories. When students in earlier sessions of the course came to him for personal conferences, he said, they most frequently asked him about the possible consequences of premarital petting and petting to climax. Realizing the pervasiveness of such forms of sexuality, he concluded that he could collect data from those conferences in order to investigate these phenomena and how they affected people after marriage. He then offered some statistical backup, taken from his case history data, for a related theme that he had been hinting at in previous lectures: "There is statistical correlation between premarital petting and effective sexual adjustment at the marriage. There is practically no correlation between premarital intercourse and ease of adjustment after marriage." He would argue in the *Female* volume that both premarital petting and intercourse eased adjustment to the sexual aspects of marriage; in this instance he may simply have been trying to avoid trouble from President Wells and enemies of the marriage course by declining to advocate full premarital intercourse. He also addressed a topic that a handful of students had inquired about—homosexuality—informing them, "Biologically, it is still part of the normal sexual picture, and the individual who suffers through the social condemnation which is a result of the branding of the phenomenon as abnormal has the most difficult sexual problem that I know of. It is a phenomenon that society will some day [face] with more objectivity." Kinsey's direct references in the lecture to individual conferences and sex histories clearly indicate their significance for his methods for teaching sexuality to IU students. The reference to homosexuality and its normalcy reinforces the fact that Kinsey was moving away from marriage per se as a subject for analysis and toward studying the whole of human sexual behavior.[66]

By the summer of 1940, Kinsey was increasingly aware of the objections of several IU faculty members to the explicit content of the course, and in particular to the idea of the normality of masturbation and homosexuality. Later in the summer, feeling pressure from IU sociologists, campus physicians, local Christian leaders, and some parents, Wells asked Kinsey to choose one of two courses of action: resign from the marriage course and continue to take the sex histories of students, or continue to teach in the marriage course but allow the IU health center to take over the individual counseling sessions. Kinsey opted to keep taking the histories, and the marriage course proceeded under the leadership of his old opponents. But he did not worry over the loss of the marriage course for long. Ten days after he sent his resignation letter to Wells, he wrote to IU graduate student Glenn Ramsey about the heterosexuality–homosexuality scale on which he was working.[67]

During his time directing the marriage course, Kinsey moved from focusing on marital sexuality to exploring and declaring the normalcy of most other forms of sexuality as well. In his final set of lectures, Kinsey also showed signs of the philosophical and scientific approach he would soon apply to the sex research for which he would gain wide renown. He built his philosophy on those elements of the marriage education and sexology literature that he had initially valued: firm statistics, clear language, a nonjudgmental attitude toward all types of behavior, an absence of religious or moral sensibility, and a desire to teach the "truth" about sex in marriage as much as anyone could. While those principles continued to guide him, he also read and thought more about sex as a cultural, social, and physical phenomenon, and kept taking sex histories as the course continued. Kinsey's 1938 marriage course lectures chastised a world where premarital contact was forbidden and marriage was fraught with problems, and subtly promoted sexual learning and experimentation before marriage. His 1940 lectures described the psychological nature of sexual experience, called for love to be removed from scientific studies of sex, pointed to the almost infinite variation in human behavior, declared the naturalness of most sexual behavior (except for abstinence), and questioned the relationship between body and mind. The honest responses that Kinsey believed he had elicited from students and others during individual sex conferences led him to think that such personal, private interviews constituted the best way to gather mass amounts of information on sexual behavior. As of 1940, Kinsey still had more questions than answers, but he now had a set of research principles at hand with which he could begin formulating those questions in a manner that satisfied his desire for scientific validity and truthfulness. By the time Kinsey's tenure as chair of the course ended in 1940, he had started to craft the ideology and theoretical framework out of which he would conduct the work that would become *Sexual Behavior in the Human Male* and *Sexual Behavior in the Human Female*.

The methods Kinsey had developed stayed with him as he began to shift research areas. Kinsey was able to apply his working methods to a humanist field in which he had developed some expertise but had not yet mastered. The possibility of producing groundbreaking research in a new field clearly fascinated him. He wrote to Voris around October 1939, "I have kept the cynipid research going, though it has slowed because of the other research problem."[68] The "other research problem" would soon absorb all of his attention. All of the skills he had learned through twenty years of researching gall wasps would become useful as he turned his "bigger side interest" into full-time sex research: research techniques and patterns of data management that were transferrable across fields; a commitment to naked-eye observation and attentiveness to the complex demands of scientific objectivity; smoothly dealing with potential controversy over teaching evolution; building networks of like-minded colleagues; engaging in the institution-building of an evolutionary science community; an intense interest in variation; an ethic of tolerance for diverse behaviors; and building on the personal resilience he developed during his years at the Bussey Institution and in the field. Kinsey never left a record that he was surprised that his research took the turn that it did. His research questions had changed, but his faith in the ability of observation and of scientific methods to answer them had not.

Kinsey moved from one area of synthesis to another in moving from textbooks to the *Male* and *Female* volumes. He approached human sex research with an existing ability to synthesize data from many sources for the production of a single text, and to organize masses of qualitative data into sortable quantitative units. While many of his peers in biology shifted to studying evolutionary processes in laboratory-created experimental animals, Kinsey shifted to studying the sexual behaviors of humans in their natural environment, with no attempt to transform who his subjects were or to modify what they did. As Edgar Anderson put it in a 1941 letter after Kinsey's move to human sexual behavior was nearly complete, "It was heartwarming to see you settling down into what I suppose will be your real life work. One would never have believed that all sides of you could have found a project big enough to need them all. . . . The monographer Kinsey, the naturalist Kinsey, the camp counsellor Kinsey all rolling into one at last and going full steam ahead."[69] And so he would.

4

Ordering Human Sexuality

> Sufficiently patient and prolonged observation may be expected to disclose, in time, various natural lines of cleavage within the total psychodynamic set-up of the individual, and when enough [people] are studied in the spirit we or our successors shall come to know what particular types of events—both inner and outer—characteristically occur in recognizable constellation and sequence patterns.
>
> —Gilbert V. Hamilton, *A Research in Marriage*

KINSEY BROUGHT INTO his worldview for studying human sexual behavior a complex mix of ideas from his entomological research, teaching life and human sciences, and teaching and organizing the marriage course. Those ideas oftentimes coexisted uneasily. He approached the more in-depth phase of human sex research using ideas and beliefs from previous research synthesized into his own view. He favored using multiple positions in a heterosexual encounter to improve the chances of orgasm for both participants, allowing children to explore their sexual bodies and feelings with little to no adult supervision, and the idea that premarital sexual experimentation was good. He thought birth control was a good idea, and that sublimation did not work. Premarital sexual experience and/or comprehensive sex education without prejudice equaled better marital sex. Regular sex equaled marital happiness. Masturbation was popular among women and nearly universal among men. On the flip side, adultery was likely, especially husbands straying from their wives. Also, rape was a problem with which he decided not to deal directly or comprehensively, and he collected and used data from admitted and convicted sex offenders. Studying human sexuality had opened a new cultural, academic, and intellectual world to Kinsey, and while some of his views on those issues solidified early in his sex research career, on others he stayed amenable to change.

Kinsey's developing perspectives on these complicated sex-related matters derived from his comprehensive reading, his contacts with individuals engaging in an increasingly wide range of sexual behaviors, and his planning to capture their wide range of experiences in forms that would allow for statistical manipulation, analysis, and publication. After Kinsey gave up the marriage course in September 1940, he was able to devote the bulk of his research time to gathering sex histories, refining his interview questionnaire and interviewing technique, and figuring out patterns in his data. In this period, he developed his interview method, the basis of its standardized and memorized questionnaire, and his methods of recording interview data. He also decided on an approach to collecting data, and a statistical method that would generate data to answer the many questions that his new research field had raised for him. He likewise adopted a form of technology, punched-card machines, that would transform the ways he manipulated data. When all of the mechanisms for the practical aspects of the research were in place, Kinsey and his growing staff could proceed more efficiently through interviews and data analysis. Even with the intellectual, mathematical, and mechanical tools for research in place by 1941, the completed *Male* volume, with its examination of more than five thousand interviews with white American men, remained seven years in the future.

This chapter outlines the three parts of Kinsey's sex research apparatus: the sex history interview content, the interview method, and method of securing interviewees; his adoption of a statistical methodology and perspective on sampling; and his enthusiastic acceptance and implementation of punched-card machinery. Each aspect of this research apparatus provided the groundwork to support the project as a whole, and together they formed the content and shape of the *Male* and *Female* volumes. Through the adoption of methodologies and machines, and trial-and-error in the first years of interviewing, Kinsey settled on methods that suited both his research questions about the nature of human sexual behavior and his research subjects. Punched-card machines made the analysis of mass data possible, and marked Kinsey's sex research as distinct from that of his interview- and letter-based forerunners in the field. The use of such labor-intensive infrastructure also made Kinsey's results difficult for other scientists to replicate. Kinsey's methods of classification and organization, and use of the accompanying punched-card machinery, separated his work from his predecessors and formed the two volumes' encyclopedia-like scope as compendia of detailed information.

The Sex History Interview

Kinsey likely found a number of his sources for the content, structure, and format of the sex history interview through his investigations of the past work that the Committee for Research in Problems of Sex (CRPS) had funded since its beginning in 1922. Many of the earliest statistical sex researchers in the United States were connected or knew each other through the CRPS, the Bureau of Social Hygiene, or mutual interests such as birth control or the current state of marriage. As Kinsey read the work of the CRPS-funded researchers, established strong ties with Robert Latou Dickinson, and applied for funding himself in 1941, he got to know the primary scholars involved in statistical sex research and became part of their intellectual world.[1] Kinsey derived ideas for the form and content of the sex history interview from Katharine Bement Davis, Dickinson, and William S. Taylor, along with Gilbert V. Hamilton, Carney Landis, and a handful of other non-CRPS-affiliated sex researchers in the 1920s and 1930s. Some of those researchers also exemplified what and how he should ask people about their personal and sex lives if he wanted truthful, accurate answers. He also decided how to manage questions about an interviewee's race and to obtain information about children's sexual behavior.

Davis and her book *Factors in the Sex Life of Twenty-Two Hundred Women* had a lasting impact on Kinsey. Davis's large sample size and thorough questionnaire provided solid proof of several points critical to the themes of Kinsey's work, regarding women's sexual desires, the relationship of sex education to marital happiness, and the neutral phrasing of the questionnaire. Kinsey was interested enough in the details of Davis's work that he asked Dickinson in mid-1946 if he or someone else had a copy of the whole questionnaire when he was making a comparative study of statistical surveys.[2] Dickinson was unable to locate an original copy, and none are extant.

Davis's survey was particularly notable for its detailed questions about homosexuality and masturbation and her approach to those topics. She reprinted some of the survey questions in those sections of the book, and they are remarkable for their lack of condemnation. Nowhere does she imply that the survey-taker should cease the behavior under consideration. For example, she referred to masturbation as "self-induced sex pleasure" that could include "sex reveries or daydreaming," "manipulation of the organs," and/or "the voluntary self-inducing of the orgasm by any means other than intercourse." Three hundred eighty-one married women and seven hundred eighty-three single women in her survey, or slightly more than half, had engaged in one of the activities described as masturbation.[3] If she had referred to masturbation using one of the negative euphemisms popular at the time, such as "self-abuse," "self-harm," or "the secret vice," probably fewer women would have stated that they participated in it.

Kinsey borrowed from and expanded on the language that Davis used in her work. Her evenhanded approach to all types of behavior affected the structure of his sex history interview for both sexes. For example, Davis's questionnaire for unmarried women included a multipart question about their sexual feelings during childhood. She asked: was there "spontaneous (physiologic) excitation of organs; pleasure in handling organs; desire for sex excitement; sex daydreams; curiosity about affairs of parents or other adults; strong attachment for boys or men; any other?" Kinsey parsed the different parts of such a question and made each of his own interview questions on child sexuality more specific:

Back before you were (age at puberty) was there self-masturbation? How old were you when it began?

How old were you the first time [you had prepubertal orgasm in masturbation]?

How old were you the first time you saw the sex organ of an adult? How did you happen to see it?

Back before you were (age at puberty), how old were you the first time there was any sex play with (girls, boys)—showing or touching the sex organs? How old were you the first time?

Was there showing sex organs? Touching sex organs? Putting a finger or object in the vagina? Mouth on sex organs? Putting the penis in the vagina?[4]

While Davis attempted to cover the whole of women's sex lives in her questionnaire, using a written form limited her ability to ask highly detailed questions without the survey-takers simply ignoring them or failing to complete the survey altogether. Kinsey and his associates could cover all of the detailed questions that they wanted in face-to-face interviews, in whichever order suited them and their interviewees the best, and it would be more difficult for interviewees to decline to answer them.

Moreover, Davis modeled sexual behavior on a ranked scale, and it was one model for Kinsey's heterosexuality–homosexuality scale.[5] Kinsey found Davis's perspective on studying homosexuality helpful, particularly her graded quantifications of same-sex behavior by life-cycle stage. Also, Davis's terminology for homosexuality, like her language in questions about masturbation, suggests no condemnation of same-sex feelings or behaviors. *H.I* was her term for "homosexual feeling without overt practices," *H.II* represented "homosexual feeling with overt practices," *H.II.c* meant "overt homosexual practices before puberty," and *I.E.R.* stood for "intense emotional relations with other women." Her numbers for each of these different behaviors were high, especially for the 1,200 unmarried women in her sample: 293 had H.I. feelings, 312 had H.II experience, 330 had H.II.c experience, and 605 had same-sex emotional relations. Davis concludes that "the phenomena described in this study are much more widespread than is generally suspected, or than most administrators are willing to admit."[6] H.I, H.II., H.II.c, and I.E.R. captured gradations in women's emotional and sexual lives that Kinsey would use in conceiving a 0–6 scale for both sexes.

Kinsey adopted many of Davis's conclusions about women's sexual behavior. He too believed that sexual instruction should be coeducational, that premarital restrictions on sexual knowledge were especially damaging to women, and, more generally, that the more people knew about sex before they engaged in it, the less personal and marital discord they would experience. Kinsey and Davis treated all types of sexual behavior on the same moral plane, not praising certain kinds of behaviors while condemning others. Equalizing the nature of many of the topics he covered was itself a bias, but one that Kinsey was more comfortable with than the behavior-specific biases in sex-related studies in other fields.

Hamilton's *A Research in Marriage* was published a few months after Davis's *Factors in the Sex Life of Twenty-Two Hundred Women* in 1929. Hamilton, like Kinsey, did his original graduate school education in animals before moving to the study of human sexual behavior. Hamilton, who had studied monkeys and baboons before training in human psychology, personally interviewed two hundred of his own patients and friends of theirs. He spoke with one hundred men and one hundred women, all under forty years old and living in New York City. He used a standard set of 370 questions, some of Davis's and some of his own, printed on typed cards. He defined orgasm in the context of intercourse as "the spasmodic, highly pleasurable feeling with which the sex act ends for both men and women."[7] Hamilton handed the cards to his subjects one at a time and transcribed their responses verbatim. He insisted on writing down the answers to the questions himself in longhand without a stenographer or electronic recording devices. Each individual, private interview took between two and thirty hours, with most taking two hours. Wardell Pomeroy remembered later that Kinsey was particularly impressed by Hamilton's description of his face-to-face interview technique.[8]

Glenn Ramsey, who conducted sex history interviews in 1942 as a member of the Institute for Sex Research (ISR) team, remembered in 1972 his own interview with Kinsey after the summer 1938 marriage course term, recalling, "I was a little tense and anxious during the first few minutes of the interview, but soon became quite relaxed in giving my history—and at the end I felt that this man really knew what he was doing." There were about 250 questions in the

Figure 4.1. Alfred C. Kinsey modeling how he took a sex history interview with Institute for Sex Research staff member Jean Brown, c. 1950. Photo by Bill Dellenback. Courtesy Kinsey Institute for Research in Sex, Gender, and Reproduction, Inc.

early version of the interview when Ramsey first encountered it. Kinsey wrote to his mentor Raymond Pearl in July 1939, "We have systematized our records and are getting something like 250 items on each student."[9] Kinsey put together data from a range of sources that he could use to craft the sex history interview in terms of the questions and of the form in which he asked them.[10]

Gradually, the interview developed into a set of 521 questions, though many heterosexual adult individuals' experiences were covered in approximately three hundred questions. In addition to basic demographic data, interviewees were asked in the course of those three hundred questions about numerous forms of sexual arousal, the timing and context for their arousal (such as a preference for nudity or clothing and a light or dark room), their coital positions, and the percentage of sexual encounters that led to orgasm. Some questions were gender-restricted, as only men were asked if they had two healthy testicles and only women were asked about menstruation. The average interview took between ninety minutes and two hours.[11] The additional two hundred questions were for men and women who had participated in homosexual activities or sadomasochism, were a part of an unspecified "special group," or who were in prison.

The interviewers needed to remember a variety of symbols for different responses as they coded them on the interview sheets, which provided space for precise answers and quantified the interviewee's level of hesitation or enthusiasm for each response. For example, both men and women were asked if they worried about masturbation. Possible answers included that they never worried, that they had a fear of mental disease, that they had a fear of physical debility, that they felt guilty, that they had unspecified worries, or simply that they did not know if they were worried. Men and women were asked in detail about their feelings and experiences about their first coitus: if they felt embarrassment, fear, shame, nervousness, regret, or guilt; if they thought it was unaesthetic or messy; if there were odd odors or smells; if it was painful; and if they were drunk. Men were asked their feelings about prostitutes, and if they preferred paid sex, sex with unmarried "companions," or their wives.[12] Studying the codebook not only shows that the interviewers needed prodigious memories but also reveals the richness of the data that was collected but did not appear in the published versions of the Kinsey Reports.

Another element of the sex history interview that did not make it into the reports—and one that illustrates Kinsey's thinking about differences between nature and culture—was a question regarding race. There were nine potential answers on the interview sheet for an interviewee's race: "white, black (dark), black (light), Mexican or any white plus [American] Indian, Cuban, South American, Puerto Rican, Oriental (Japanese, Chinese, Philippine, etc.), [and] other." In the *Kinsey Interview Kit*, the racial category has the following explanation: "We are mainly after the cultural aspects of race rather than the biological aspects. If a person has Negro blood, but passes as a white, he would be called W, with explanatory note. If a person is part Indian and part N, but is thought of as N by his associates, he would be called N (with also indication as part Indian). . . . If a person is foreign-born and raised, the foreign country is put in parentheses after the race."[13] Kinsey himself probably did not write that explanation of how to measure an interviewee's race; Paul Gebhard or Pomeroy likely wrote it when the interview kit was prepared for internal use at the Kinsey Institute in the mid-1980s. Nonetheless, they reflect Kinsey's training regarding what he and the other interviewers asked for and how they coded their responses. The explanation of race in the *Interview Kit* supports the idea that Kinsey thought that "biological" and "cultural" concepts of race were separate. The interviewee's perception of racial identity and his or her understanding of how others perceived his or her racial identity were both more important than any biological meaning of race. Kinsey had first articulated that biological/cultural separation of racial meaning in *Edible Wild Plants of Eastern North America*, and he brought forward that understanding of race into the reports.[14]

Another data-gathering issue that Kinsey had to manage was that of children's sexual behavior, which remains one of the most controversial aspects of his research. Kinsey gathered data on child sexual behavior from three sources. First, all adult interviewees were asked about their prepubertal sexual experiences, including their prepubertal play with same-sex and opposite-sex children and same-sex and opposite-sex adults. Questions included what techniques the interviewee used in his/her play, whether he/she was the active or passive participant (or both) in the play, whether he/she was aroused, what relationship the interviewee had with the other participant, and if the interviewee continued the same type of behavior into adolescence or adulthood.[15] Though adults' earliest memories may not have been exact, adults could at least give rough estimates of when they had their earliest sexual feelings and experiences.

Second, Pomeroy wrote that the confessed pedophile Kenneth S. Green's experiences, which included six hundred preadolescent boys and two hundred preadolescent girls, were "the basis for a fair part of Chapter Five in the *Male* volume, concerning child sexuality."[16] Green also kept his own records of his experiences with adolescents, other adults, animals, and incestuous sex with seventeen relatives. The most notorious of the tables in the chapter "Early Sexual Growth and Activity" was table thirty-four, "Examples of Multiple Orgasm in Pre-Adolescent Males," which lists the number of multiple orgasms that boys from five months through fourteen years of age had over a single period of up to twenty-four hours. Other tables list preadolescent eroticism and orgasm generally, the speed of preadolescent orgasm, and the number of boys experiencing multiple orgasms, "based on a small and select group of boys. Not typical of the experience, but suggestive of the capacities of pre-adolescent boys in general."[17] Kinsey and Pomeroy used Green's records and remembrances and the other two sources to support the idea that men were sexual across their lifetimes.

The third source of data about children's sexual behavior was from children themselves. Children were interviewed only if at least one of their parents was present, and if the parent had already given his or her own sexual history. Kinsey wrote in the *Male* volume regarding interviewing children less than eight years of age that an interview "becomes a social session involving participation in the child's ordinary activities. . . . Tucked into these activities are questions that give information on the child's sexual background."[18] Interviews with children involved discussions of the child's activities around bedtime, drawing pictures, and the child's interactions with other children and his/her response to those interactions. Kinsey described the process to the gynecologist Sophia Kleegman not long after the *Male* volume was published:

> Our usual procedure is for us to speak to the parents or parents and teachers first, and to go after histories from the parents and teachers. We do not take any history from a child until at least one of his parents has given us a history

and has consented to our getting a history from their child. We never take the history of a child under seven unless at least one of the parents is present at the interview. We prefer to take the histories of the children under seven in their own homes, or in some other place, such as a school, with which they are familiar. . . . This children's material is precious. It is giving us more incite [sic] into the patterns of behavior than we even anticipated we could get.[19]

Together, those three sources on children gave him enough data for him to describe observations of both child orgasmic behavior and adult orgasmic behavior as being on a continuum from mild convulsions to intense release to pain or fright as orgasm approached. For small children and babies who were too young to be interviewed, or for most people who could not remember that far back in their childhoods, observers were the only source of data on child orgasms. Orgasm, despite differences in degree, was a usable measure not only to compare the sexual activities of adult men and women but also to determine the sexual activity of children. If Kinsey was going to use orgasm as the standard measure of outlet, he wanted to use it for all of his subjects, even though the data from many of the youngest subjects came from those who hurt them. Kinsey may have established that orgasm happened in the same way for children and adults, but that knowledge came at a price to those who had been abused for it. As long as Kinsey believed a source was telling the truth, he chose not to discriminate among sources regarding what crimes, sexual or otherwise, they had committed in the past.

Kinsey also needed language and a measurement device for his specific questions about the physiology of sexual behavior in adults. Davis, Hamilton, and the psychiatrist Carney Landis all used orgasm as a primary measurement for their adult studies. Davis defined orgasm as "a convulsive contraction of the muscles of the interior sex organs, followed by a definite relaxation." Hamilton, as stated above, referred to orgasm as "the spasmodic, highly pleasurable feeling" with which a sex act ended, though of course many erotic acts had no physiological conclusion of that sort. Landis used the word "outlet," which may also have inspired Kinsey's use of the same word to define orgasm. Each author described orgasm in nearly identical language, even though they had quite different perspectives on what orgasm meant in people's lives.[20] If studies as different as the ones Kinsey read agreed on its definition, perhaps it was indeed the most scientific measure available. While orgasm was hardly a measure closed to interpretation for determining the whole of an individual's sexuality, it was more precise and neutral than the emotion-related measures, such as "happiness" or "satisfaction," that other sex researchers used.[21] Kinsey's decision to use orgasm as a measure of sexual behavior did not make his analysis objective, but it was an attempt to obtain records of sexual behavior with some clarity as to the nature of one of the physiological events involved.[22]

In addition to choosing questions and a measurement device, Kinsey also had to decide on a satisfactory method of sampling the American population. Probability sampling, which would have provided him with a proportionate representation of individuals across the US demographic, was not in wide use in the late 1930s and early 1940s, and the time, energy, and funds needed to conduct a sex history interview survey in that fashion would have made it almost impossible to carry out. Sarah E. Igo puts Kinsey's decision not to use probability sampling in the broader context of statistics and mass polling. She points out that leaders in the statistical establishment at the time had not yet fixed a standard sampling technique. Statisticians were slowly coalescing around probability sampling as their favored method, but that method had its own problems.[23] Kinsey was not alone in looking for and using alternate methods of sampling from several different statisticians. Consequently, Kinsey needed a method that he could enact and could justify later in his writing. He came upon the idea of 100 percent group sampling, a method of sampling one of the earliest sex researchers in the United States had used. That researcher was Max Joseph Exner, a member of the American Social Hygiene Association who conducted the first published quantitative study of sexual behavior among groups of college men in 1913. The purpose of his study was to gather information on sexual behavior "in order to direct the sex education movement more scientifically and effectively." When he traveled to a new college, he gave a lecture on social hygiene to a classroom full of college-aged men before handing out surveys. "The lectures served to bring the students into cordial relation to the speaker and to gain their confidence," he wrote.[24] The lecture also served to inform the men of his stance on sexual matters, and of what kinds of answers he might expect. Exner's questionnaire included requests for basic statistical information, such as the age at which an individual began any kind of sexual practice, and questions about sex education that required moral evaluation, such as "Indicate in what way this information (concerning the subject of sex) was good or bad for you."[25] Exner focused on a population with similar attributes and gathered data from both "Eastern" and "Western" colleges: men of the same age and education level, likely the same social and income level, and probably all white. The keys to obtaining sex histories from Kinsey's subjects would be similar to those that Exner and later, other social hygienists, had already set in place decades before: anonymity, rapport, and an appeal to participate in the creation of scientific knowledge that would be useful to the public. Exner also had a data-gathering technique called "the one hundred percent sample" that strongly shaped how Kinsey collected his own sample.[26]

While Exner used a questionnaire, and the language of his questions was full of moral judgment against anything but abstinence prior to marriage, he also made an important observation about the percentage of surveys that the young men returned. He got the most surveys back when no members of a group, usually in a college hygiene class, were allowed to leave the room until they had all finished a survey. Indeed, his presence before and probably during survey completion was crucial to ensuring that the men filled it out. Exner's ability to go from a 20–25 percent return rate (when teachers allowed men to opt out of taking the survey) to a 100 percent return rate simply by controlling traffic flow in a room likely impressed Kinsey. Kinsey decided to visit groups—fraternities, civic clubs, religious organizations, and the like, and convinced the leaders of those groups to give their sex histories. With their encouragement and some peer pressure, he was able to obtain the histories of the whole group. Kinsey favored this method, which he also called one hundred percent sampling, but it was not always possible. So occasionally Kinsey and the ISR team would take the histories only of the willing portion of the group. They named that method "partial group sampling."[27]

Kinsey gained much from his additional reading across sex-related surveys, including those of Davis, Hamilton, and Exner. He centered his reading on works that used face-to-face or survey interview data to make conclusions, had a relatively large sample size, treated sexuality with a minimum of moral, religious, and ethical framing, used precise and scientific language, and did not shy away from nonmarital and solitary types of sex. As of early 1941 he had an interview form and a set of research principles at hand with which he could begin formulating those questions in a manner that satisfied his desire for scientific validity and truthfulness. He realized that with 521 possible data points for each interviewee, he needed sophisticated statistical techniques to analyze his gathered material. In this period, when Kinsey was thinking through problems of his new sex research agenda, he also met Raymond Pearl, a good friend of William Morton Wheeler, who influenced Kinsey regarding his decisions to use biometrics and 100 percent sampling in his sex research.

Raymond Pearl and the Patten Lectures

Kinsey's sex research program required knowledge and use of an interdisciplinary range of intellectual devices. The biologist and biometrician Raymond Pearl inspired Kinsey's thinking on variation in the natural world, the possibilities of human-animal comparison, and studying sexual behavior without reproduction. Kinsey knew of him by 1935, as he praised Pearl in his 1935 lecture to a peer group of professors, "Biological Aspects of Some Social Problems."[28] Pearl's lectures on the IU campus in the fall of 1938—published six years after Pearl's death in 1940—confirmed, or perhaps inspired, Kinsey's views on "variation" and "normalcy" in natural populations. When Pearl visited Bloomington in 1938, he became an influence on Kinsey in terms of ideas about normalcy, variation in behavior, the collection of sexual material beyond interviews, and focusing attention on behavior rather than reproduction.

Pearl, a professor at Johns Hopkins University, had a varied research career, and his work explored multiple scholarly areas, including entomology, genetics, population studies, and statistics.[29] He was not content to explore his own research interests, as he also founded two peer-reviewed academic biology journals, *Quarterly Review of Biology* and *Human Biology*. When Pearl visited Bloomington, he had recently returned from London, where he had delivered the Clark lectures at the University of London. Those lectures on comparative reproduction statistics across the United States and the Western world were later published as *The Natural History of Population*, which would affect Kinsey's views on measuring the natural world.[30]

The Patten lectures were an opportunity for students and faculty at IU to interact with and to learn from an established academic who might otherwise never visit the rural state university. An alumnus of IU, Will Patten, gave a grant to the university to pay for a senior scholar in any field to be in residence at the university for a period of six to eight weeks once a year. Pearl was the first, and his 1938 diary records the events of his trip to Bloomington. Pearl arrived on September 27 and stayed through November 12. He gave five Patten lectures, one of the marriage course lectures, one talk to the science honors fraternity Sigma Xi on "Biological Factors in Human Fertility," and attended one event that the sociology department gave in his honor in that period of time. He also attended five of the weekly Sunday evening musicales that Kinsey held at his home on First Street, and the two men also met at other social events and at least four other times alone. Kinsey and Pearl met at least eighteen times during Pearl's residency in Bloomington.[31]

Pearl's discussion of the likenesses between humans and mammals in his first two Patten lectures, "The Unique Mammal" and "The Unique Mammal (Continued)," likely resonated most with Kinsey. Pearl began the first lecture by stating that "man's firm biological anchorage to his solid mammalian mooring is the greatest source of his strength in the cosmic scheme of things." Pearl then became more specific about the similarities between humans and animals. Their physiology and behavior were alike, so scientists could study animal physiology and behavior to learn about that of humans: "If the essentials of man's physiology are identical with those of mammalian physiology in general, it is reasonably to be expected that much of *his behavior* will be found upon analysis to be like that of other mammals in similar or equivalent situations."[32] Kinsey's language comparing humans and mammals in both volumes of the Kinsey Reports echoes Pearl's in the Patten lectures.

Since "a large part of human behavior is just animal behavior," Pearl reasoned, one could assume that animal and human sexual behaviors were much alike as well. Further, Pearl argued, "human courtship behavior stems from deep mammalian roots. The proof of this cannot be undertaken in a public lecture, but must be left for the seminar and laboratory." Here Pearl dodged the potentially thorny problem of discussing his ideas on human sexual behavior explicitly. As Pearl had given his marriage course lecture a week before he gave his first Patten lecture, and had attended at least one other marriage course lecture (Fowler Harper's), the subject of comparative human and animal "courtship" behavior and mating would have been on his mind for the Patten lectures. He was also in the process of editing his Clark lectures for publication, so it is unsurprising that he was ruminating on other comparisons as well. Such human–animal comparisons would also be important for Kinsey as he was thinking about gathering, organizing, and analyzing his growing amount of sex history data.[33] Kinsey's work itself may have challenged or stimulated Pearl, so perhaps Pearl was using the Patten lectures to make a public, albeit veiled, statement about the importance of the marriage course and of Kinsey's collection of sex histories.

As Pearl affirmed the likenesses of human and animal behavior, he likewise stressed the uniqueness of each individual. Individuals, Pearl stated, all differed, but they differed in small amounts clustered around a "center of variation." Indeed, "such concepts about natural phenomena as 'normal,' 'typical' or 'truth' itself are merely the expressions of majority opinions so widely concurrent and congruous as to be practically universal."[34] In other words, Pearl thought that just because some behaviors were common in the world, this did not mean that less common behaviors were abnormal. Those who behaved in rarer ways than the majority did not deserve condemnation either. "There is no absolute or transcendental normality or type," Pearl continued. "This is a point well to remember when we get into the more complicated realism of human biology such as sociality, political behavior, and the like."[35] For Pearl, if "normal" behavior was an artificial category that researchers created, so was "abnormal" behavior, and he implied that human society should not stigmatize either kind of behavior.

Pearl created a table, "The Relative Variability of Normal Adult Human Beings," to illustrate for his listeners his point about the great amount of human variation in the world. The table delineates six types of characteristics weighted for the conceivable amount of variation in each. Pearl intended to create a mathematical formula to describe the total extent of human variation. How he created his formula is not clear, but he weighted the relative variability of human behavior at 135.52 times the relative variability of more stable factors such as body size and shape. The behavior category includes such measurements as "smoking, use of alcoholic beverages, frequency of coitus, etc." That formula does not appear in either of Pearl's two later books, *Natural History of Population* and the third edition of his textbook *Introduction to Medical Biometry and Statistics*, so he did not explore its application further in print.[36] However Pearl came up with the formula, it may have had an impact on Kinsey's thinking about his gathering and measurements of sexual behavior. Since for Pearl, human behavior had the most possibilities for variation out of any other aspect of human nature, Kinsey perhaps felt reaffirmed in his decision to focus on sexual behavior for his next major scientific project. Further, Pearl's repeated references to coitus and sexual behavior affirmed that human sexual behavior was a topic of scientific, and not only moral, legal, or medical interest, and they lent support to Kinsey's own focus on sexual behavior, from marital coitus to many other types.

Kinsey did not record his reaction to Pearl's lectures, but Pearl's insistence on the variety of human behavior, and the fact that behavior differed more than any other physical or human attribute, surely resonated with him. It also confirmed the importance of the variation he was discovering in his sex history interviews. Pearl's ideas that behavior differed most of any human attribute, and that sexual behavior was alike in humans and in mammals, could coalesce in Kinsey's mind for the support of his own views on approaching sex research. Pearl's focus in his last three lectures on reproduction, the primary topic of his next book, resounded with Kinsey in another way.[37] Perhaps Kinsey thought that Pearl's extensive work on reproduction and population growth covered those topics enough that he could borrow some ideas from Pearl and analyze sexual behavior without reference to reproduction.

Cornelia Christenson, in her 1971 biography of Kinsey, quotes a letter that Kinsey wrote about his meetings with Pearl in Bloomington circa the fall of 1938. Kinsey affirms that Pearl encouraged him to continue to gather large numbers of sex histories to make up for his lack of mathematical ability. Kinsey wrote, "He points out that statistical theory is largely a substitute for adequate data. All of this encourages me greatly. We are scheduled for a long session together on the subject of sampling, pretty soon." Robert Kroc, who taught in the marriage course, recalled in an interview with James H. Jones that Kinsey told him about Pearl's encouragement of Kinsey's work on the sex history interviews.[38]

That "long session together on the subject of sampling" would have a marked influence on the mathematics of Kinsey's work. Pearl's ideas on sampling and statistics, which structured the third and final edition of Pearl's textbook, *Introduction to Medical Biometry and Statistics* (1940), became the base text that Kinsey used to frame his study of sexual behavior mathematically, to solidify his method of statistical analysis, to figure out what interpretations he could make from his gathered statistics, and to determine the number of cases he needed for his sample. Pearl's book also provided him with some ideas for justifying his decision to use 100 percent sampling instead of random sampling. Another statistical textbook, George W. Snedecor's *Statistical Methods Applied to Experiments in Agri-*

culture and Biology, was Kinsey's source for arguing that his method for ultimate group sampling—used with 100 percent and partial group sampling—was the best way to obtain sex histories from a variety of people without random or probability sampling.[39]

Raymond Pearl was an advocate of biometry, a form of statistics developed in the late nineteenth century to analyze the natural world numerically. Pearl's simple statement that mathematical methods could be used to measure the natural world accurately had a strong impact on Kinsey. Kinsey had determinedly not applied statistical techniques to his study of evolution in gall wasps, and he in fact criticized those who did as using artificial methods to understand the natural world. Pearl claimed that biologists could use biometrics in ways that would not falsely manipulate their data. Instead, "with the development of knowledge and of an appropriate technic [sic] eventually any natural phenomenon which can be observed can also be quantitatively measured." Biometry is specifically focused on "statistics derived from living things," and is designed "to help the scientist to draw correct conclusions from his facts, and to solve problems constantly arising in his work, which he cannot possibly hope to solve correctly without such methods."[40] As Kinsey was confronted with the complexities of the data he was collecting from interviewees, he may have been grateful to discover a form of statistical analysis that did not compromise his views that mathematics separated the scientist from the realities of his data and of the natural world. Additionally, Pearl's description of biometry and his methodology would have been relatively easy for Kinsey, as a nonprofessional mathematician, to understand and adopt as his own as compared to theoretical statistical works.

The methods of biometry were not from standard statistical methods at the time. The difference was in the subjects of the two fields' analysis. Biometric textbooks like Pearl's, or statistical textbooks focused on agriculture and biology like Snedecor's, were designed for doctors, public health workers, and biologists who needed assistance with the mathematics of their work. Exercises in Pearl's textbook used data from his own studies of jaundice symptoms, corn kernels, and mortality rates; those in Snedecor's contained problems with pig production, seeds, and the weight gain of hogs.[41] Learning statistics with practical examples would have reassured Kinsey that statistics could be used for human studies and not be an artificial imposition on his interview data. Compared to earthy examples like those, the work of statisticians who focused on abstract theory or formulas would have seemed impractical to him. While Kinsey was not a mathematician, he recognized the necessity of using math to understand trends in his human sexuality data, and learning statistics through biometrics fit his analytical needs.

Pearl outlines two uses for biometrics in his textbook, both of which were important to Kinsey. The first is "furnishing a method . . . of describing a *group* in terms of the group's attributes, rather than in terms of the attributes of the individuals which compose the group."[42] Biometrics could show the composition of groups, their typical condition, and their degrees of individual diversity. That use of statistics is only precise when scientists limit their conclusions to the group under study, and when they do not extrapolate beyond the limits of their data. "But when we endeavor to predict from that particular group to other groups or individuals or to conditions in general, our results are no longer precise, but inferential," Pearl continues.[43] Kinsey wrestled with the extent to which he could use his results for deductions about individuals and groups beyond his immediate sample. Periodically throughout the *Male* volume, Kinsey repeats that his conclusions concern only the 163 groups for whom he felt that he had adequate data. His tendency toward making conclusions from his data that some critics saw as overreaching would draw the criticism of statisticians and others after the *Male* volume was published.

The second use of biometrics was "a wholly different aspect of the statistical method." Pearl argues that this aspect is to be "used for the purpose of predicting or estimating the probable or the approximate condition in the *individual* from a statistical examination of the condition in the mass or the group." Kinsey adopted that use of biometrics to predict outward from the group under study for the statistical tables that appear at the end of the *Male* volume. Pearl was careful to state that scientists should be wary of how much they infer about causation. Statistics, "being only a descriptive method, tells us nothing *directly* about the causes involved in the determination of any events or phenomena.... It may be of great aid ... in helping to arrive at such knowledge, but alone and of itself it cannot directly furnish knowledge of causes of individual events."[44] Kinsey probably extrapolated from Pearl's second use of biometrics the idea for the *Male* volume's statistical tables. He and his team provided group statistics on amounts of sexual behavior for the use of clinicians and policy makers, and they used the tables to make inferences about clients or the applicability of policies as they saw fit. An ASA review team and many others would later criticize Kinsey's conflation of the two uses of biometrics, seeing them as unnecessary and too speculative.[45]

Pearl's assertion that large numbers tend to increase the accuracy of a scientist's results would have confirmed Kinsey's existing beliefs about the importance of a large sample size. Kinsey had favored obtaining large numbers of specimens as a measure of precision from the beginning of his work on gall wasps forward, and his view of sex histories was no different. Pearl acknowledges "the notion that there is a special virtue ... in such knowledge as is reached by the examination of large numbers of cases" but cautions against depending only on large numbers as a determination of correctness. Pearl affirms that it is necessary to balance the average scientist's interest in obtaining large samples with "the concepts of averages, approximation, and probability," and Kinsey found ways to manage those other three requirements for accuracy in his own fashion.[46] Pearl states later in the book, however, that "the reliability or trustworthiness of any conclusion is in some way a function of the number of cases upon which it is based."[47] That is true in a statistical sense, as two types of statistical errors—probable errors and standard errors—become smaller as the size of the sample increases. Even if there are fewer errors in the statistical calculations, though, that does not reduce likely problems with the reliability of the data in the original sample or consequent errors of interpretation. Kinsey certainly believed that "the examination of large numbers of cases" made his sample more accurate than those of others. Kinsey aimed for both large numbers and accuracy, but not for random sampling. His lack of a random sample led him to create and to defend the accuracy of his specific form of group sampling.

For all of Kinsey's attention to Pearl's two purposes of biometrics, he set aside Pearl's concerns that probability and randomness of a sample population improved the chance that the sample would reflect behavioral realities. Kinsey may have downplayed Pearl's point about the importance of random sampling, but he took Pearl at his word that a larger sample would improve his accuracy. Kinsey put together a method that combined the work of Pearl and Snedecor: a large overall sample plus analysis of—and a focus on—specific small groups for accuracy. Kinsey created his own method of representative sampling from Pearl's and Snedecor's work and satisfied his desire for breadth and examination of multiple characteristics. In Kinsey's reading of Pearl, he found no way around the fact that contemporary statistics required a random sample. He was convinced, however, that the refusal rate for random sampling for his project would be so high as to make collecting sex histories untenable. In other words, "Kinsey favored the stratified sampling of predetermined subgroups of the population over the representative sampling of samples drawn at random from the population as a whole."[48] He instead read Pearl and Snedecor for how he could use statistics to justify his decision to use his 100 percent sampling method, and to convince readers that his nonrandom sampling method was just as valid and accurate as random sampling.

While Pearl helped Kinsey with the theoretical background of biometrics, Kinsey found Snedecor's work as he was deciding how to explain his statistical decision-making. Kinsey needed to show his readers, especially those familiar with mathematical methods, that his sampling was valid even if it was not random. He signed and dated a copy of the fourth edition of Snedecor's *Statistical Methods Applied to Experiments in Agriculture and Biology* on May 2, 1946, as he was intensively preparing the *Male* volume.[49] Snedecor's section on reducing the error involved in nonrandom sampling caught Kinsey's attention. Snedecor, like Pearl, states that obtaining large samples is one method of reducing statistical error. Another common method that Snedecor discusses is "subdividing the aggregates into sub-populations or *strata* of similar individuals." If a researcher found that random sampling was impossible—usually for reasons of time or money—that person could aim to select individuals that represented the total of possible outcomes or behaviors in a group. Snedecor called "stratified sampling" a method in which the researcher would find nonrandom samples that matched certain criteria for a group, along with samples that completely represented the variation in the group. Kinsey wrote the phrase "stratified sampling" alongside the description of the method in his copy of Snedecor's book. However, there was no way to know in advance the potential for variation of a group when sexual behavior was the focus of the research, so Kinsey and his team risked the error of meeting one of Snedecor's requirements for stratified sampling (selecting samples that met certain visible criteria, like age or marital status) and not the other (ensuring that the behavioral variation of the group was completely represented). "The efficacy of this method of stratification depends on foreknowledge of the behavior of experimental material, and this can only be gained by experience," Snedecor wrote.[50] Combining Snedecor's and Pearl's methods may have worked best for situations where the researcher could have a more complete idea of the likely variation before choosing samples for the groups, but Kinsey and his team, especially in the early stages of their research, could not predict what an interviewee would say in advance of his or her interview.

For Kinsey, finding Snedecor's description and justification of stratified sampling helped him explain why he saw the breakdown of his large sample into smaller groups with specific characteristics as valid. If the researcher could predict that a sample with certain characteristics would behave a certain way, selecting a small sample on the basis of its homogeneity (i.e., religion, sex, age, race, and marital status of a sex history interviewee) could reduce error. The ability to define his reasoning for 100 percent and partial group sampling still did not solve the problem of his lack of random sampling, but fully outlining his method in the *Male* volume would be a way for Kinsey to show his readers how he tried to reduce sampling error with his chosen methods. According to Snedecor, the best researchers have to find a balance between keeping the group under discussion homogenous in its surface characteristics, and ensuring that the variation endemic to the group is also represented. Confirming that a non-random sample yields both "an unbiased estimation of the population mean" and "an unbiased estimate of variance as well as the mean" is Snedecor's challenge to would-be statisticians. Kinsey wrote the phrase "representative sampling" next to Snedecor's explanation of balancing group specificity and variation in his copy.[51]

Neither Snedecor nor Pearl was totally comfortable with nonrandom sampling, as even the most objectively minded researcher could introduce bias into a selected sample, but they acknowledged that it is perhaps the best option for some research projects. Pearl's benchmark for statistical accuracy is not random sampling but the elimination of bias or judgment on the part of the researcher in selecting the sample. As Pearl wrote in *Medical Biometry and Statistics*: "In all statistical endeavor the idea ever to be striven for is to devise such mechanistic schemes or systems of sampling as will automatically exclude all necessity, and still more importantly, all opportunity for the exercise of judgment, fairness, or any other similar affective operation of the mind of the sampler."[52] One hundred percent group sampling was the best "mechanistic scheme or system" for Kinsey to remove as much bias as he could from his sample and to obtain histories from persons willing to give them. He did not want to waste time asking random potential interviewees for their sexual histories when they were quite likely to say no. His method for collecting histories may not have met the standards of most contemporary statisticians, but it satisfied the criteria of the statistician whom Kinsey trusted to teach him. Perhaps as Kinsey was writing the *Male* volume, he found solace in the fact that at least a couple of statisticians understood that there possibly were good reasons to do nonrandom sampling. He learned from them that he could make an argument for nonrandom sampling based on the limits of his and his team's time and energy. The refusal rate for using 100 percent group sampling would be a lot lower than the rate for random sampling, and he decided to put together a complex justification for a nonrandom sample rather than risk the time and expense involved in gathering a random sample.[53]

Pearl and Snedecor both had multiple influences on Kinsey. Pearl's and Snedecor's brands of biometrics, geared toward analysis of life and human science data, would strongly shape Kinsey's statistical theory. Pearl also spent much time with Kinsey at a crucial moment in the development of his sex history data-gathering technique; he was an intellectual omnivore like Wheeler; and he and Wheeler were good friends throughout their adult lives.[54] Kinsey kept in touch with Pearl until the latter's death in 1940, and he was excited to share the news of his sex history collecting with Pearl in July 1939. "We have over 350 complete histories so far," Kinsey wrote. "This should be of some scientific interest someday. In the meanwhile it is giving us the basis for answering a great many of the questions that the students bring us."[55] Kinsey's reference to the marriage course in that letter shows the impact that the course and his related research were having on the trajectory of his thought. His one-on-one work with Pearl, the Patten lectures, and Snedecor's textbook reinforced for him that it was typical, even expected, that life scientists would explore a wide range of topics over the course of their careers. Most importantly, Pearl provided him with statistical knowledge, a sampling strategy, and academic justifications for both.

The Mechanization of Sexual Data

The *Male* and *Female* volumes were literary heirs to a tradition of scientific human sex research beginning with Richard von Krafft-Ebing and Sigmund Freud in late nineteenth-century Austria.[56] The inclusion of long narrative histories of individual patients characterizes the work of each author, whether they were physicians, psychiatrists, biologists, or sociologists. They sometimes included some simple percentage calculations if they used a questionnaire. The subjects of the interviews were normally patients of the author, or in some cases author and subject knew each other only through exchanges of letters. The author would then use patient data and that correspondence to make broad generalizations about the sexuality of the study population and often about humanity generally. Their books were intended for use by fellow clinicians and professionals, not for the public. Sometimes the original texts would be in Latin or in Greek, in order to deter the casual reader from coming across salacious material. Kinsey's publications, with their focus on statistics derived from large samples, were a marked break from the norms of sex research literature. As the Kinsey Reports elevated quantitative methods in sex research, they downgraded what had been the most heavily used form of human sex research—the aggregation of individual sex histories from patients or therapy clients.

Kinsey was collecting and managing much larger quantities of data than his predecessors, and he intended his published research to reach a larger audience. In order to process and display this increasingly large mass of data, he chose to incorporate machines into his research project. The data collected and analyzed in the Kinsey Reports could not have been processed or analyzed without machines. The use of machines marked a clear difference between the qualitative, nontechnological sex research that psychologists, psychiatrists, and physicians performed and Kinsey's new quantitative, technologically oriented vision of sexology as an interdisciplinary science. Outlining the process of data transfer from interview sheet through punched-card machines shows that the means of processing interview data shaped Kinsey's ability to create and to support models of sexual behavior such as the 0–6 scale.

Punched-card machines (a.k.a. Hollerith machines) were in wide use in businesses, governments, and universities when Kinsey first envisioned using them in late 1940 and early 1941. His decision to adopt them for sex research was part of the "continuous spread in usage of the punched card machinery" worldwide from the 1920s through the 1950s. Also, as "the punched card was the basis for the most advanced information technology from the 1920s to the Second World War," Kinsey wanted his sexual science methodology and machinery to be as organized, stable, and accurate as those of his peers in the life and human sciences.[57]

The basic unit of any mechanical tabulating system was the punched card itself. Rectangular holes in the cards represented data points made at preset arranged positions on the card. The perforations could be "sorted, counted, and tabulated by a series of machines, automatically, as often and in as many different ways as desired by the operator."[58] The three separate machines needed to handle basic operations were the key punch, the tabulator (a.k.a. accounting machine), and the sorter.[59] The key punch was "used to transfer the initial data of a problem from the manuscript to the punched cards." A blank card would be inserted into the punch, and the machine operator would select where the punching knives would make holes in the cards. The speed of the punching operation depended on the quickness of the operator, the clarity of the manuscript, and the amount of punching needed. The author of the first guidebook for using punched-card machines estimated that for an "extensive job" an experienced operator could process between 200 and 125 cards per hour. If multiple cards needed to be punched at the same time, data could be transferred from one card to another by wiring the key punch to punch subsequent cards the same way as previous ones.[60]

The sorter automatically sorted the cards into groups according to the punches in any chosen column of cards. The cards were placed in the "card hopper" at one end of the sorter, the desired data points were selected by adjusting the movable sorting brush, and cards would then sort into any of thirteen receptacles. Kinsey had the sorter type 75, which had a device that counted the cards in each receptacle.[61] Kinsey's type 405 tabulator was "essentially a high-speed adding machine of large capacity [that read] the numbers to be added from holes in punched cards." As a card passed through the feed mechanism with its eighty brushes (one for each column on a standard card), the adding wheels revolved and tallied punches until the feed was completed. The numbers added from counting card perforations from the desired data points then were shown on dials or printed on paper. Working the tabulating machines required real physical labor. The cards had to be entered into and removed from each machine and hand-carried to the next. Keeping cards organized after tabulating and sorting also had to be done by hand, usually by filing the cards in large cabinets.

All parts of a punched-card operation had function and meaning, and all were automated for efficiency and flexibility. Kinsey would use most of those functions in his statistical calculations. For example, all the cards with punches for extramarital intercourse could be sorted with those for a certain age group and social level, and the staff could then calculate the rate of extramarital intercourse for that age group and social level. However, the use of punched-card machines for data processing required the data's creator to conceive of its organization beforehand in such a way that it could be transferred onto punched cards and subsequently ordered and sorted mechanically. Punched-card ma-

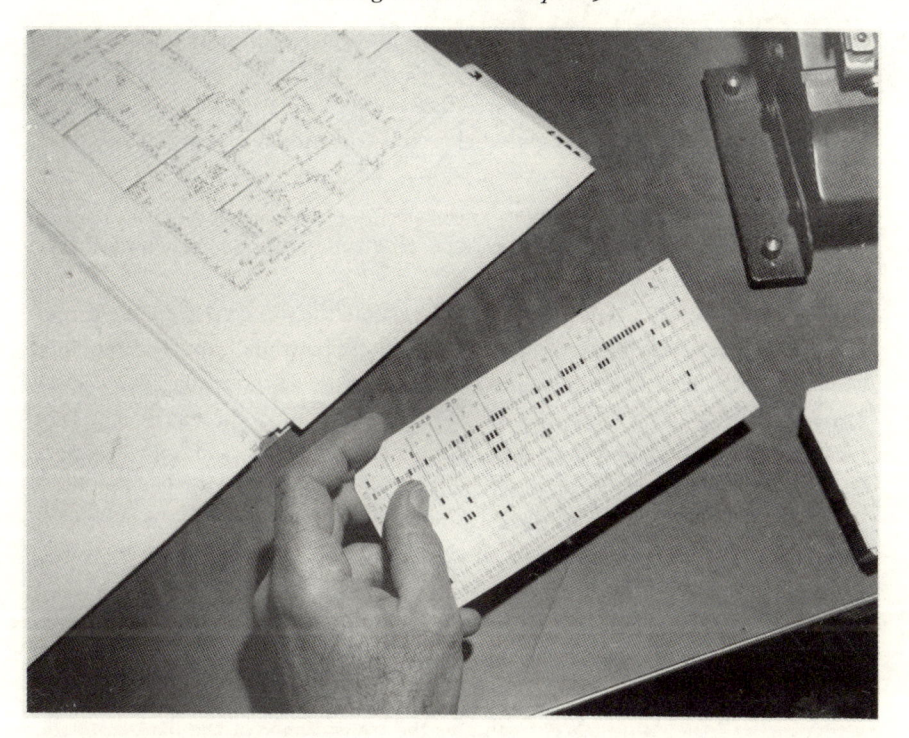

Figure 4.2. A sex history interview sheet compared with a punched card, c. 1950. Photo by Bill Dellenback. Courtesy Kinsey Institute for Research in Sex, Gender, and Reproduction, Inc.

chines could not be easily adapted for a given task, as "their use required a high use of standardization and formalization of the tasks to be processed, which, in turn, made greater demands on the user organization than did competing technologies."[62] Kinsey took on the time-consuming challenge of preorganizing interview data so that it could be moved directly from the interview sheet to the punched card. Kinsey's interview code sheet, as outlined below, shows that he developed its format so as to transfer the data easily to punched cards.

The sex history interview sheet that Kinsey and his fellow interviewers used is organized methodically, with each square designed to hold the answer to a question, written on the sheet in code. The answer sheet is divided into seven vertical columns, each of which is then subdivided into three or four thematic subheadings with varying numbers of spaces beneath them for answers. The interview sheet is ordered in columns and rows of rectangular units, and thus the data contained on the sheet was transferable to similar columns and rows on cards for machine coding. For example, the first column includes the subheadings "Health," "Marriage," and "Anatomy," and the second column includes the subheadings "Erotic Arousal," "Family Bg. [Background]," and "Sex Ed. [Education]."[63] If the interview and answers were routine, all of the answers could fit in the boxes designated on a single interview sheet. If not, the interviewer wrote answers by hand in the space below the grid and on a second blank sheet of paper. The sex histories of prisoners, especially prisoners with sex offenses, often required the extra space. As Kinsey wrote in the *Female* volume, "it is a mistake to believe that standard questions fed through diverse human machines can bring standard answers."[64]

Kinsey, Pomeroy, Clyde Martin, Gebhard, and the three other men entrusted with the whole set of questions memorized both them and the accompanying code to record the answers. The unwritten questions and the answer sheet codes protected interviewees' identities, speeded the recording of their answers, made it easier for the interviewer to see if he missed a question, and kept outsiders from being able to crack the code. Machine coding added another level of identity protection and secrecy to the ISR's work. Robert Yerkes, chair of Kinsey's primary funder, the Rockefeller Foundation–backed Committee for Research in Problems of Sex, and thus a highly important figure in Kinsey's world, had his sex history taken on a visit to the ISR in early December 1942 and was impressed by the layers of confidentiality Kinsey had instilled in his data-collection process.[65] "Coding at the time of interview serves several functions," Kinsey wrote in the *Male* volume. "It facilitates the transference of the data from the original record sheet to punched cards for statistical analyses." The questioner always showed the interviewee the mysteriously coded sheet, so interviewees would be reassured that their contributions would stay anonymous and confidential. The relationship between the interview sheet and the punched cards in the *Male* volume is clear: "In each block [of the interview sheet], the available symbols are sufficient in number to designate all of the categories into which the particular data will be classified during subsequent analyses."[66]

Kinsey's first grant from the CRPS made it possible for Kinsey to put his data-organizing plans into action: he used the grant to purchase a calculator and to rent tabulating equipment.[67] Once the machines appeared, Kinsey and Martin began to transfer the interview data, which had previously existed only on the handwritten interview sheets, into punched-card data. Pomeroy, who joined the project in 1943, noted that "both Martin and Kinsey made the mistakes beginners usually make, and it took some time to effect the transfer properly."[68] Certain features of punched-card machinery—particularly the anonymity created by aggregating the sex history data on cards and its ability to process large quantities of data quickly—were especially appealing to Kinsey as he began to work more closely with the punched-card machines. In March 1941, he wrote excitedly to Glenn Ramsey, the IU education graduate student and former participant in the marriage course:

Figure 4.3. Clyde E. Martin, coauthor with Alfred C. Kinsey of *Sexual Behavior in the Human Male* (1948) and *Sexual Behavior in the Human Female* (1953), using a punched-card sorter to organize sex research data, c. 1950. Photo by Bill Dellenback. Courtesy Kinsey Institute for Research in Sex, Gender, and Reproduction, Inc.

> I immediately see that it will save us endless hours of work in analyzing our data. It will be possible for us to run correlations of an indefinite number of factors, at least eighty in any one problem—a thing which is utterly impossible by any hand calculation. Wherever we have a tabulation of more than perhaps five hundred cases, wherever we have a problem of figuring frequencies, and wherever we are interested in correlations, it pays to transfer our data to punch cards and get the answers by machine. We have just completed our first set of punched cards which covers the heterosexual-homosexual formulae for the entire lifetime of the individual, and such correlative items as age of adolescence to frequency of outlet per week, etc. I had despaired of ever analyzing these formulae by hand techniques. The machine will do it at the rate of 400 cards per minute. This new equipment is a godsend to our particular problem.[69]

Each individual sex history took ISR staff eighty minutes to punch into thirteen cards, which were then organized into thematic sets, but once those tasks were completed, the data contained on the cards could be manipulated into nearly any desired configuration. By March 1942, Kinsey and Martin had spent approximately fifteen hundred hours punching fifteen thousand cards with data from twenty-eight hundred interviews, of which eleven hundred interviews were from April 1941 to March 1942 alone. As Kinsey wrote in his March 1942 report, five of the thirteen sets of cards were prepared, so "the Hollerith cards for the study on Frequency of Sexual Outlet are now ready for the machines, and the manuscript on that part of the study should be completed by the end of the summer."[70]

Once the staff had transferred the interview data onto standard IBM 5081 cards (twelve rows, eighty columns each) with the key punch, they used the tabulator and sorter to correlate various combinations of data points with each other—such as age, educational level, and homosexual behavior to orgasm, to name one example. Kinsey gives credit in his April 1948 report to Clyde Martin specifically for taking charge of data manipulation, saying that "Mr. Martin has done some unusual things in analyzing data on these machines."[71] As Bruno Latour has stated about the longevity of data organization, "All these charts, tables, and trajectories are conveniently at hand and combinable at will, no matter whether they are twenty centuries old or a day old." With every new entry on a sex history interview sheet, "new relations of distance and proximity, new neighbourhoods, new families are devised. . . . At each translation of traces onto a new one something is gained."[72]

Kinsey clearly placed a high value on his tabulating machines, as they increased the efficiency and productivity of his research. Yet he did not discount the economic investment and physical labor involved in ensuring that the machines performed the work that he needed them to do. The tabulating machines made complex computations and correlations of the Institute's constantly growing data sets possible, but with a high premium on the staff's time. As Kinsey and his staff collected approximately 18,000 histories over the period of the project, and since each history filled thirteen cards, they punched around 234,000 total cards between 1941 and 1956. If they never got faster at the card punching than the eighty-minutes-per-history figure given in 1942, punching 18,000 histories took 24,000 hours. Once punched, however, all of that data was readily maneuverable, though Gebhard remembered that "in those days we lacked computers and our card sorters were slow. A relatively simple table could easily take a full day or two of sorting."[73] Those tables made statistical data visible in graphic form to accompany the many pages of descriptive text in the *Male* and *Female* volumes.

The tabulating machines also provided a method for Kinsey and his staff to answer common questions about sexual behavior according to age. They figured out a method to represent visually how many individuals had engaged in a

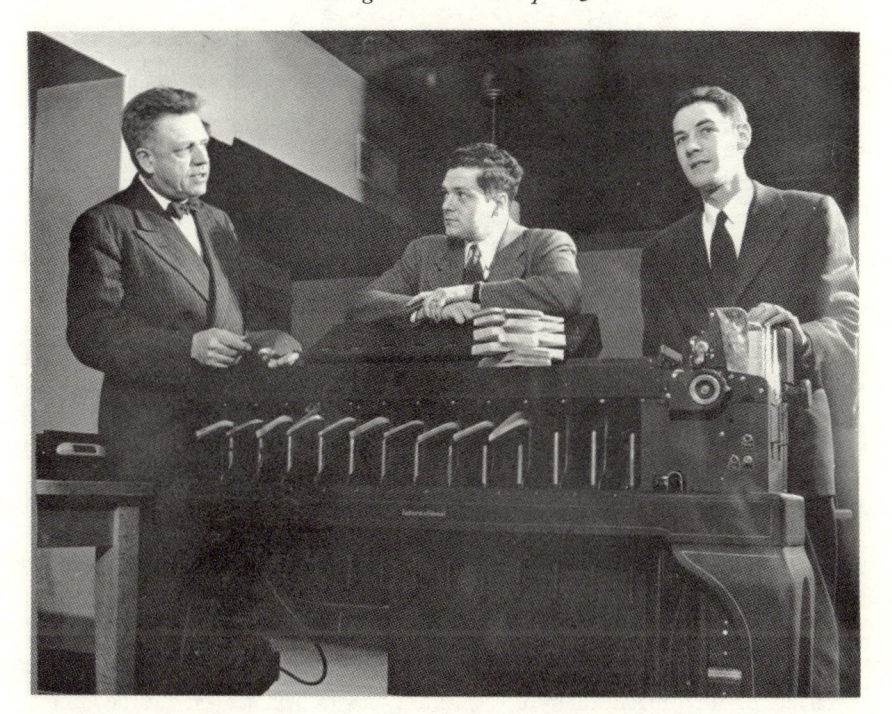

Figure 4.4. Alfred C. Kinsey, Wardell B. Pomeroy, and Clyde E. Martin standing around a punched-card sorter full of cards, 1947. Kinsey holds a punched card in his hands. Photo by Bill Dellenback. Courtesy Kinsey Institute for Research in Sex, Gender, and Reproduction, Inc.

certain behavior by a certain age, which they called an accumulative incidence curve. Its derivation, Kinsey wrote, "was first worked out for a small sample by a hand manipulation of 1058 actual history sheets, adding them to piles as each individual became eligible, withdrawing them as each individual became ineligible for experience." Such a procedure was clearly time-consuming, but ultimately it helped the staff to set up the tabulator to correlate age at the first incidence of each of the behaviors under study, and then to correlate those two data points with others, such as educational or social level. "It took some time to devise a procedure for Hollerith machine manipulation of punch cards on the problem, but a remarkably simple set-up has now been arrived at," he concludes.[74] Kinsey deeply embedded tabulating machines across his research tasks, and he wrote to Ramsey again in March 1941, detailing another triumph of data manipulation. "Martin and I are doing the most stupendous piece of work we have undertaken yet," he wrote with amazement. "We are transferring something around six hundred items from the histories to punched cards to be run through our statistical machine. It is a very slow process, but when we get this set of cards done, we will have all the data that will have a bearing on marital adjustment ready for instant analysis."[75] Kinsey's focus on marital adjustment is unsurprising, as he was writing to Ramsey only six months after the IU marriage course ended. One reason the IU administration forced his resignation from the course was his contention that premarital sexual experimentation would speed sexual adjustments after marriage.[76] He was curious if quantitative evidence could support his contention, and by the time he wrote the *Male* volume in 1946–47, he found that it did.

The inescapable presence of machine-produced data in the Kinsey Reports emphasizes the importance of quantitative methods in the human sciences. The *Male* and *Female* volumes integrate machine-processed data and qualitative material together. The use of large quantities of machine-processed data and the conclusions drawn from them marked the sharpest break between Kinsey and his predecessors in sex research. Although no sex researchers followed Kinsey's lead in using punched-card machines to organize and to classify their data, they began to use machines in their research in other ways. The most famous sex researchers of the 1960s, William H. Masters and Virginia E. Johnson, made one machine, the penis-camera, the centerpiece of their research program, and used a variety of other machines to provide additional information about the human body as a sexual entity.[77] Though punched-card machines eventually became obsolete, Kinsey's work signaled the new centrality of machines to human sex research.

When Kinsey began a research program designed to amass data about human sexual behavior, he faced the possibility that there was a nearly limitless amount of potential information he could collect about the subject. As a scientist with a predilection for mass collecting and well-honed classificatory skills, he partitioned an interviewee's life and sexual experience into 521 data points that he could keep apart or could reassemble as outlines and relationships among them appeared. He asked thousands of individuals questions that no other researcher was asking, and interviewed people whose behaviors fit some expected outward social patterns but challenged others. The *Male* volume was a taxonomist's attempt to make meaning out of great quantities of data, and to sketch a representation of the complexities of human sexual experience. Kinsey needed statistical methods and punched-card machines to perform the mathematical functions that made mass data comprehensible but were too time-consuming to be done by hand. Machines made the Kinsey Reports a reality, but they also obscured the nuances embedded in "the complicated interchanges between interviewers and subjects."[78]

After more than six years of interviewing thousands of people and punching their sex history data onto cards, weeks of cross-country travel, copious and comprehensive translation and reading of secondary sources, and long hours of writing, revising, and editing the manuscript of *Sexual Behavior in the Human Male*, Kinsey and his fellow ISR staff members would see their hard work in print in the first week of January 1948. Kinsey had found his publisher, the medical publishing house W. B. Saunders, after impressing the house's president during a well-received lecture in Philadelphia.[79] The fruit of their labor, which runs to 804 pages, showcases the product of their large, intensive, and time-consuming project on the sexual behavior of men and their partners with extensive arrays of charts, graphs, images, and textual descriptions. The *Male* volume would manifest the results of the lead author's lifelong interest in the power of classification and data organization to reveal new knowledge about the natural world.

5

The Taxonomy and Classification of Human Sexuality

> The present volume is a taxonomic study of the frequencies and sources of sexual outlet among American males.
> —Alfred C. Kinsey, Wardell B. Pomeroy, and Paul H. Gebhard,
> *Sexual Behavior in the Human Male*

SEXUAL BEHAVIOR IN THE HUMAN MALE reflects Kinsey's synthesis of material from a wide assortment of disciplines coupled with the data from 5,300 male histories. The book shows the results of his comprehensive interview and research formula: his attempts to fill in the "great gaps in exact knowledge" of human sexual behavior.[1] His shift to using machine-organized data and quantitative methodology signaled a shift in the most popular and lately dependable tools of sex research. Alfred Emerson, who had known Kinsey since the early 1940s during their mutual work for the Society for the Study of Speciation, praised the aims of the *Male* volume three weeks after it was published. Kinsey, he wrote, was "opening up a new approach to human social science that is of the utmost importance." Kinsey concurred: "I too hope that the social sciences will have guts enough to try this method on some of their other problems. It does take time and lots of work, and it remains to be seen whether they think it is worth it."[2]

Kinsey's classificatory scheme for the whole *Male* volume—first examining sexual behavior through social categories, then through categories of sexual behavior themselves—was grounded in what he viewed as the most objective aspects of previous sex researchers' work. He divided the book into three parts: the first, "History and Method," concerns the study's overall methodology; the second, "Factors Affecting Sexual Outlet," contains nine chapters examining the possible reasons behind sexual behaviors; and the third, "Sources of Sexual Outlet," contains nine chapters on different types of sexual contact plus the clinical tables. In the first chapter of the *Male* volume, Kinsey walks his readers through his research philosophy, the work of his statistical sex research predecessors, and his assertion of the legitimacy of measuring natural phenomena quantitatively. He describes in detail the mechanisms that he and his team used to create rapport with subjects before and during their individual interview sessions in the chapter on interviewing. He outlines his 100 percent and partial group sampling methods and statistical techniques to show readers his methods and to convince them of their veracity and appropriateness to his work. As social level was the only topic that merited one chapter on its behavioral effects and nearly a whole second chapter on its manifestations over time, the effects of Kinsey's background readings in those two chapters becomes particularly evident. While men were active throughout their life cycles, they showed great variation in terms of the types of behaviors and arousals. Nonetheless, the explanation and inclusion of the statistical tables reveal Kinsey's position that men's behavioral variation was best kept inside class boundaries.

This chapter investigates the *Male* volume with a focus on how Kinsey's classification structure affected the data and his conclusions about it. It focuses in particular on the introductory chapters, where Kinsey outlines his method, its importance, and its break from past sex research; the chapters on social level and the stability of social patterns, which show how Kinsey applied his classification model; and the chapter on homosexuality, which illustrates most clearly the intellectual fruits of the use of machine-generated data on his thinking about the organization of knowledge around sexual behavior. By ordering individuals on a scale of 0–6 instead of a heterosexual-bisexual-homosexual triad model (or even a binary heterosexual–homosexual model), Kinsey shifted the modern conversation about sexual behavior toward a form of classification that re-envisioned a world in which multiple varieties and combinations of sexual desire, behavior, and fantasy would be culturally, scientifically, and politically normal.

A Taxonomy of Human Behavior

The first chapter contains Kinsey's explication of his taxonomic methods and their importance to sexual science and to the human sciences more generally. Taxonomic methods, he argues, are applicable to the human sciences with some difficulty but much reward. As far as his own research is concerned, "the transfer from insect to human material is not illogical, for it has been a transfer of a method that may be applied to the study of any variable population, in any field."[3] He details how he collected the sample of sex histories that he used, for if the sampling were poor, the data could be meaningless. As sampling was particularly challenging when approaching individuals for private or sensitive data, Kinsey explains his 100 percent and partial group sampling methods in detail, and the reasons why he decided on them. It was important to collect enough data so that the full range of possible variation was evident in the sample. "Each investigator must know the general order of the variation that may occur in the material with which he works, see to it that the sample is well spread through the range of variation, and learn through some pragmatic means the general order of the sample size that will begin to represent the whole of the universe that is being sampled."[4] Furthermore, collecting a representative sample was the mark of a professional, skilled taxonomist, and Kinsey wanted to demonstrate that properly done taxonomy could apply to human behavior studies. The aim, after all, of collecting a good sample was to ensure that the scientist and the reader could generalize about the data presented to other persons with similar characteristics:

> If individuals are collected in a fashion which eliminates all bias in their choosing, and in a fashion which includes material from every type of habitat and from the whole range of the species, it should be possible to secure a sample which, after measurement and classification, will indicate the frequency with which each type of variant occurs in each local population, or in the species as a whole. . . . If the sample is adequate, the generalizations should apply not only to the individuals which were actually measured, but to those which were never collected and which were never measured at all.[5]

The taxonomist needed to demonstrate to which individuals the research applied and to which it did not. Kinsey decided to analyze his white and nonwhite data separately, and to focus on white men in the *Male* volume, in order to prevent critics from attributing his data on "deviance" to nonwhites and his "normal" data to whites.[6] Kinsey was well aware that the *Male* volume did not have adequate samples of all the male Americans that he wanted to study eventually. But for smaller subsets of the published sample, he felt confident about his data's generalizability.

Further, Kinsey believed, if the sample was adequate, the taxonomist could claim the right to analyze the data collected and to postulate reasons for, and connections between, its characteristics. After all, "the taxonomist finds the different backgrounds where they are already established in nature and, if his investigation is accurate, can reason as the experimentalist does about causal factors."[7] Kinsey distinguished himself from scientists who viewed themselves as experimentalists, as they controlled the environment of subjects and analyzed or isolated factors to see which ones made a difference. Kinsey, however, as a taxonomist, prided himself on collecting data "already established in nature," as "nature" was where he believed scientists would find the operational truths of the world. As in his days as an entomologist and would-be evolutionary scientist, Kinsey continued to downplay the usefulness of applied scientific methods in favor of original research.

Kinsey intended, in the introductory chapters of the *Male* volume, to delineate his method of gathering data, and to describe how that method elevated and distinguished his work as a whole from the work of those who came before him. Contemporary critics strongly contested Kinsey's claim that he had the right to interpret his own data and disagreed with him about most of his methods and established parameters. They disputed the idea that he was dispassionately and objectively collecting, organizing, and interpreting, let alone capturing the nature of human behavior.[8] Kinsey, perhaps anticipating some of those criticisms, turns in the last third of the first chapter to a critique of his immediate predecessors as a foil for demonstrating the accuracy of his methods and sample compared to theirs. His discussion of elements like objectivity, the interview, attracting subjects, and rapport shows how he absorbed the work of others into his own.

The first chapter ends with a description of the available published statistical sex research in several fields, which Kinsey found mostly inadequate. One of their common problems, he wrote, was their sampling method: "All of the studies taken together did not begin to provide a sample of such size and so distributed as a taxonomist would demand in studying a plant or animal species." His sample, of course, was the largest, besting Robert Latou Dickinson and Lura Beam's four thousand cases by more than 20 percent.[9] Second, "obvious confusions of moral values, philosophic theory, and the scientific fact" plagued many of the other studies that came before his. For example, many suffered from problems of classification due to artificial value structures that supported some behaviors while condemning others, like masturbation—"far outrun[ning] scientific determinations of the objective fact," Kinsey concludes.[10]

Despite all that Kinsey had actually gained from reading sex research, his desire to distinguish himself led him to contrast its flaws with his successes and originality. After Kinsey critiques his predecessors, he concludes that "there seemed ample opportunity for making a scientifically sounder study of human sex behavior." As Kinsey uses the word "sound*er*" rather than "sound," he indicates that the other scholarship he examined and used had some value. Approaching sex via his interdisciplinary methodology, he thought, offered "scientifically sounder" data to the professional and lay public than any that had come before.[11] There is only one place in the bibliographical review where Kinsey specifically acknowledges the impact of another researcher, describing "the most notable aspect" of Max Joseph Exner's sampling technique as "the '100 per cent' sample secured by getting records from every one of the 673 males in a series of groups."[12] One hundred percent sampling as the way to acquire the most accurate and complete samples of groups, however, was Kinsey's own idea.

Kinsey describes the intellectual tools that he brought to the sex history interview in detail. Kinsey and his team needed to "master" data-gathering techniques—such as the face-to-face interviewing techniques in Hamilton's *A Research in Marriage*—in order to use human personalities and memories as "instruments." Their work also required the best classification and organizational schemes of others. Kinsey affirms, "[The] complexities of such a study constituted a test of the capacities of our science." Kinsey may not have insecticized his subject's sex histories in the *Male* volume, but he did transform qualitative human narratives into quantitative data sets for statistical analysis. He avows that natural phenomena can be measured using statistics. Previous studies, from Lilburn Merrill's on delinquent boys to Carney Landis's on mentally handicapped women, showed that researchers could persuade a "wide variety" of subjects to participate.[13] Kinsey's own experience gaining the confidence of diverse populations, along with his reading of "the more recently published research" provided him with "a considerable basis for deciding what should be included in a sex history."[14]

A clear written definition of orgasm was part of the method that Kinsey had established for his sex research, along with the tenor and composition of the history interview. He defines orgasm, which he called "outlet" in the reports, as "a sudden release which produces local spasms or more extensive or all-consuming convulsions." Significantly, he adds that "the moment of sudden release is the point commonly recognized among biologists as orgasm," giving the term the weight of approval of earlier sexual scientists, including Davis, Hamilton, and Landis. Though Davis and Landis derived their working conception of orgasm from studying women, Hamilton used his definition for both sexes. Kinsey's use of the event as an equivalent measure for both sexes indicates that he too thought male and female orgasms were alike enough that they were equal measures of sexual arousal. Kinsey spends little time explaining why he chose to focus on orgasm as a measurement, other than that "the event happens to both men and women alike and other types of emotion are not as clearly measurable or countable."[15] Orgasms were indeed countable in women, but, as he details in the *Female* volume, they often played a different role in their sexual lives.

Kinsey divides the sources of outlet into six types: masturbation, nocturnal emissions, petting, heterosexual intercourse, homosexual relations, and relations with other species. The sum of orgasms from all of those sources "constitutes the individual's total sexual outlet." Even though almost all adult sexual contacts resulted in some sort of arousal, those contacts were so variable that they were not easily measurable. Thus, "for the sake of achieving some precision in analysis, the present discussion of outlet is confined to those instances of sexual activity which culminate in orgasm."[16] Measuring the same phenomenon across each type of sexual behavior for each individual also ensured consistency in data collection, especially when analyzing the whole of that person's experience. Kinsey's account of statistics reveals the mechanisms he created for sampling populations, analyzing his data, and defending his choices.

Kinsey limits the generalizations offered in the *Male* volume to "163 groups on which data is given," despite the inclusive-sounding nature of its title. He does not claim the universal applicability of his work. "It is disconcerting to realize what scant bases there have been for over-all statements that have been made in this field," he states at the end of chapter one. "The present study is designed as a first step in the accumulation of a body of scientific fact that may provide the basis for sounder generalizations about the sexual behavior of certain groups and, some day, even of our American population as a whole."[17] Kinsey uses the world sound*er* again to reiterate that previous research contained valid material. But he also clearly wanted to mark his own achievement as the best. The *Male* volume itself, with its extensive bibliography and in-page citations, is a synthetic accumulation of existing sexual facts, opinions, theories, and studies in addition to the case history data that Kinsey and his team collected. After Kinsey acknowledges the influence of those twenty-three earlier studies, he explains how they factored into the creation of his sex history interview and the gathering of his own original data.

Kinsey first reviews the statistical and sampling techniques of other researchers in chapter one before he specifies his own in chapter three. He confined the literature review proper to twenty-three American studies that he considered scientific, statistical, based on case histories and series of some size, and "involving a systematic coverage of approximately the same items on each subject." Kinsey comments on each of those studies in alphabetical order. He created a chart that compares the studies according to such factors as the professional background of the author, the nature of the interview, the size of the sample, the extent to which the results could be generalized, and perhaps most importantly, the percentage of their questions that were covered in the *Male* volume. The *Male* volume itself is included at the bottom of the chart. None of the other surveys, except for Glenn Ramsey's, contained more than 25 percent of the possible questions on the interview sheet for men. The possible extension of validity to other groups is listed as "none" or some highly specialized group, such as "Mid West urban white males of Jr. H.S. age" (Glenn Ramsey, who worked in Illinois) or "Male college students" (Kenneth Martin Peterson, who worked in Colorado). At the bottom of the chart, the *Male* volume is listed as

"521 = 100%"—it not only contained three times as many possible questions as the others, it also could be validly extended to the "163 groups on which data are given."[18] He proceeds to express his own statistical method in detail.

Kinsey used a discussion of Raymond Pearl's biometry to confirm that statistics could truly be useful and accurate for the study of human sexual behavior. They would also be helpful for inferring, if only cautiously, the behavior of like members of a group who were not members of the original sample under consideration. "It is, precisely," Kinsey wrote, "the function of a population analysis *to help in the understanding of particular individuals by showing their relation to the remainder of the group.*"[19] Such knowledge, he imagined, would assist clinicians who could use it to determine if a man's sexual behavior matched the patterns of his group and of the broader population. As Kinsey would later show, social class was a significant factor in determining the behavior of many men. Knowing how men fit into their social group could dissuade clinicians or those in power (such as social workers, police officers, or judges) from giving advice or punishment based on their own class background when it was different from that of the communities they served. Knowledge of group mores, particularly as they pertained to class, would help forestall the class conflict over sexual behavior that took place at the time in governmental, medical, and legal settings, and which existed in large part because of different class standards for sexual behavior between lawmakers and the public at large.

Kinsey's discussion of variation in his group sampling echoes Pearl's argument for how scientists could most accurately measure and represent the variation of any population under study. He also reiterates points about variation that he made earlier in his textbooks and gall wasp work about the near-endless amount of variation in the natural world. He states what a researcher would need to know about the range of variation in a group so as to determine its parameters: "In order to understand any group it is necessary . . . to secure a sample of such size as will show the full range of the variation of the group, and show the frequency with which each variant occurs in the group." Kinsey clarifies his intention to discover patterns in groups and not make assumptions about the whole population: "The chief concern of the present study is an understanding of the sexual behavior of each segment of the population, and that it is only secondarily concerned with generalizations for the population as a whole."[20] Nonetheless, it was easy to read the book as if it applied to all American men.

Further, Kinsey argues that his data-gathering process ensured that the variation in behavior that he and the other interviewers found among their subjects was detailed and accurate. He was also proud of how the coded interview sheet facilitated rapid data processing. In his description of the sex history interview, Kinsey highlights the advantages of coding for machine processing. Coding at the time of the interview "facilitates the transference of the data from the original record sheet to punched cards for statistical analyses," and it was "of supreme importance in conserving space, making it possible to put the whole of the basic history on a single sheet." Coding and punching cards from data taken in longhand, as most researchers had recorded sex histories previously, were "slow and sometimes well-nigh impossible procedures." The coding within the boxes on the interview sheet mapped the possibilities of responses for each question, and each possible response was also given its own designation for accuracy. Thus "it is necessary to anticipate the whole array of possible classifications, including those that lie beyond usual experience."[21] If the data was inaccurate in the first place, even the most elaborate coding scheme could not make it usable.

In the publication of the codes for the sex history interview sheet in 1985, the codes for each box measured hesitations, confidence, and shadings of responses. For example, double checks indicated an extremely positive response to a question, and double Xs indicated an extremely negative response. The interviewer would put quotation marks around answers that he thought the respondent answered untruthfully. He would then write his idea for the accurate answer and place it in square brackets. The interviewer put an exclamation point by any answer he considered both true and unusual.[22] Such finely grained detail showed subjects' reluctance or willingness to discuss certain behaviors, and could add to estimates of behavior if the interviewer suspected an affirmative answer even if the subject was unwilling to confirm it. The more nuanced the response, the more accurate the data became, and the more precisely it represented variation within the study population.

Kinsey follows those statements with an elaborate description of the mathematical procedures that he and his team used to determine how many sex histories were needed to show the range of behaviors for a group with six parameters: sex, marital status, age, education level, religion, and level of religious devotion (active or inactive). Kinsey specifies the breakdown of data, and reiterates his earlier statement that the volume concerns itself primarily with specific groups of individuals and only secondarily with sexual diversity in the whole white American male population. One of the purposes of the study was to examine the role of different social and biological factors on sexual behavior, he wrote, because "scientifically and social[ly] it is of greatest importance to understand *why* populations differ."[23]

Kinsey's decision to put together Pearl's and George Snedecor's statistical methodologies is clear in the *Male* volume's chapter on "Statistical Problems." Kinsey based his rationalization for using statistical methods of his own devising on the complexity of the data under study. As Kinsey had up to 521 data points for 5,300 men to consider, and he knew that his lack of random sampling would likely be problematic for readers, he justifies his intricate schema for data analysis by citing Pearl's data showing that behavior varies the most of any human characteristic. "Behavior characters vary even more than physiologic characters, and these in turn vary more than morphologic characters." That sentiment echoes Pearl's idea that differences between humans in behavior are exponentially greater than other kinds.[24] Kinsey uses Pearl and Snedecor as bases for explaining why he collected a large number of interviews and then broke them down into what he called "ultimate groups."

To secure accurate data on each possible group, regardless of its rarity in the general population, it was necessary to have "more or less equal samples from each of the ultimate groups." Those groups amounted to three hundred or four hundred individuals each. An ultimate group would have five or six traits in common, such as: single, male, 16–20 years old, graduated high school but not college, and inactive Protestant. Ideally, the characteristics of interviewees as a group would already be proportionate to the characteristics of the US population as a whole. Kinsey's method of "stratified sampling" sought equal numbers of histories from all of the groups under study, regardless of how many members of those groups there were in the whole American population. However, it was not "feasible to stand on a street corner, tap every tenth individual on the shoulder, and command him to contribute a full and frankly honest sex history." The team instead weighted its data against US Census data for a given group in order to determine how frequently a behavior happened among its members nationwide.[25]

Some members of a group still refused to give histories despite the time given to building rapport with groups of potential interviewees and the peer pressure used on holdouts. So Kinsey thought that "a combination of hundred percent sampling and controlled partial sampling seems the best that can be done." Because interviewees, deliberately or not, covered up some aspects of their behavior, had trouble estimating amounts of irregular behavior, did not keep sex diaries, or may simply have forgotten that they had participated in something, "there can be no great precision to the calculations." However, Kinsey claims the ability to pinpoint where the errors might be and to determine their approximate size by retesting. Kinsey also includes a discussion of how he, Pomeroy, Martin, and the other short-term interviewers, Ramsey and Vincent Nowlis, compared with each other, and how consistent he was in his own findings across time. He repeats that "it should . . . be recognized that the data are probably fair approximations of the fact."[26] Kinsey asks his readers to trust his methods and numbers in the service of one of his larger purposes—to show which behaviors were common and which were rare in particular groups of American males.

Kinsey turns to the findings in the data that he and his colleagues had gathered after explaining and defending *Sexual Behavior in the Human Male*'s method for accuracy. His methods uncovered for human sexual behavior what they had uncovered with the gall wasps: an abundance of variation, which they then broke down into smaller groups of data to analyze. He anchors his conclusions by placing them in the context of the interdisciplinary research that he had previously synthesized. As he entered the diverse literatures that had their own collective understandings of sexual matters, he made his work not only part of intellectual conversations about sexual behavior but also part of those about the place of social class in shaping behavior, measurements of the natural world, and the interpretive reach of his and alternative research methodologies.

Social Class and Variation among Men

Chapters five through twenty-two of the *Male* volume describe and interpret the data from 5,300 of the sex history interviews with men that Kinsey and his team had gathered since 1938. He found, perhaps unsurprisingly, that much of his data supported one of the general scientific principles that he had espoused throughout his career: that variation in a single species is almost limitless. Through Kinsey's eyes, looking at variations in human behavior reveals a world of complexity and interrelationship that is not easily explainable. Variations in human behavior provided a marked challenge for Kinsey, as there were numerous correlations and possibilities for determining the relationship of one trait to another.

Kinsey found such striking data among variations in terms of social class that he decided to write two chapters focused on class instead of one, as had been devoted to other variables such as age, religion, and marital status. The first chapter on class parallels the others in terms of exploring the impact of one variable on sexual behavior. The second explores how sexual patterns changed (or did not) over time with a focus on class mores. Kinsey speculates on the reasons for the occurrence of class changes in sexual behavior, especially among children and adolescents in the second chapter as well. He found that there was no single "American pattern" but "scores of patterns," and that "social categories are realities in our Anglo-American culture."[27]

Class structured the initial approach that Kinsey made when asking an individual for his or her sex history. The primary distinctions between Kinsey's approaches were class-oriented, but each approach "is based on the measure of altruism that is to be found—if one knows how to find it—in nearly all men."[28] Interested researchers could approach lower-class men in bars or on the street, middle-class men at club meetings, and upper-class men at professional functions. Other surveys avoided class-related problems by restricting their research largely to one class or social level, but Kinsey and his team deliberately sought out interviewees at all social levels for the sake of comparison. The chapters on class were also an opportunity for Kinsey to reflect on the relative influence of nature and culture on behavior.

The opening of Kinsey's first chapter on class affirms his belief that a combination of forces created an individual's sexual makeup. At this point in his thinking, he surmised that internal bodily processes structured individual behavior to an extent, but an individual's psychological experiences trumped their biological inheritance. "The psychological bases of behavior," Kinsey argues, are "even more important than the biological heritage and requirements." Kinsey found that social level was one psychological element that clearly structured men's sexual behavior. Social class mores, Kinsey says, "may control behavior as effectively as though they were physical restraints."[29] He found that peer, family, and general social pressure exudes more power on men's sexual behavior than does biological structure. Kinsey wanted to make his correlations between social class and sexual behavior accurate, and he needed well-founded descriptions of the different social classes and occupational levels to order his findings clearly. Kinsey derived those accounts of class and occupational level from scholarship on class and on social mobility.[30]

W. Lloyd Warner and Paul S. Lunt of the University of Chicago, Pitirim Sorokin of the University of Minnesota, and several other social science researchers provided Kinsey with the instruments he needed to classify social level in his sample population. Sorokin used a six-level scale in his class analysis and ordered his classes according to occupation. F. Stuart Chapin, a colleague of Sorokin's at the University of Minnesota, also used a six-level scale to describe the social level of homes in four American cities. Education was especially important when tracking social mobility, as it was often higher education, in addition to interaction with a peer group of a different class, that moved a man into a higher social class. The equation of educational level with social level was in circulation among educators as well as sociologists. If sociologists and other academics were using education level as a proxy for social level, Kinsey felt justified in doing so too.[31]

Warner and Lunt, in *Social Life of a Modern Community*, popularly known as the first "Yankee City" study, state that their view of social class was based on ranking socially superior or inferior positions to one another. Children, in their view, are "born into the same status system as their parents." At the same time, "a system of classes . . . provides by its own values for movement up and down the social ladder."[32] They argue that class differences are evident in children's peer groups, and that peers exert a powerful influence on children of the lower classes in particular, theories that Kinsey tested in his investigations.

Kinsey cites Chapin and two Yankee City studies as his sources for the class aspect of his analytical structure. He uses both an occupational level scale and an education scale and explains the correlations and distinctions between them. There are ten classes in his occupational scale from one to ten: dependents (one), underworld (two), day labor (three), semi-skilled labor (four), skilled labor (five), lower white collar (six), upper white collar (seven), professional (eight), business executive (nine), and extremely wealthy (ten).[33] Kinsey focuses his analysis on the seven classes between day labor and professional group inclusive. He separates discussion of children into a standalone chapter ("Early Sexual Growth and Activity"), states that he did not collect enough data on men of the underworld (i.e., pimps, mobsters, and money launderers) to include them, and collapses the top three categories into one. Thus his working occupational structure has seven working levels instead of ten. He then uses three educational levels: elementary education (completing eighth grade or less), secondary education (completing at least the ninth but no more than the twelfth grade), and college education and above (completing at least one semester of college).

Occupational level and education level had related purposes in determining a man's social level. Barriers between adjacent occupational classes are ill-defined, but they "probably show a closer correlation with the intangible realities of social organization." Occupational level was a more accurate mark of the differences in how people tended to view each other in everyday life. However, educational levels often corresponded to social class, given that a specific level of education was often required for jobs. Other researchers found that educational level "proved to be the simplest and best-defined means for recognizing social levels."[34] Kinsey mostly uses educational level designations throughout the *Male* volume and turns to occupational level only when he wants a more finely grained analysis.

Kinsey discovered strong education level dissimilarities in his sample. College-educated men masturbated and petted the most of the three levels, and grade- and high school–educated men had higher levels of premarital intercourse than the college men. The grade- and high school–educated men also had intercourse with prostitutes more often than the college men did. Kinsey argues that those differences were due to upper-class emphasis on technical virginity before marriage and the lower class's disinterest in restricting intercourse for those who desired it.[35] Occupational level four (semi-skilled labor) had the highest mobility rates up and down the class scale, so perhaps class instability had something to do with high levels of sexual activity. Those disparities piqued his interest in seeing if class differences changed or stayed the same if a man moved up or down social levels during his lifetime. Class divergence in social mobility was also an opportunity for Kinsey to explore how well his statistical method explained complex variables over time. There was no gradual correlation between level of education and level of homosexual behavior in subjects' sexual history, however. "The active incidence figures are the highest among single males of the high school level. In the late teens nearly every other male of this level (41%) is having some homosexual contact, and between the ages of 26 and 30 it is had by 46 per cent of the group," Kinsey wrote. That same age group also had the highest number of single men who rated higher than X, 0, or 1 on the 0–6 scale of any other educational level.[36] The chapter "Stability of

Sexual Patterns" is also Kinsey's attempt to establish sexual science as having a potentially broader applicability to and impact on established areas and methods of social science. Though the chapter is not wholly devoted to class, the section on changes in class-related sexual behavior patterns is its largest. Warner's and Lunt's work made clear that class structured observable behavior in everyday life; Kinsey confirms from his data that class affected sexual behavior as well. He observes in the opening of that chapter that "there is, obviously, a considerable shifting of occupational classes and social position in our society and it is of interest to know how sexual patterns are affected when such changes occur in social classes."[37]

Kinsey divided his whole sample for the "Stability of Sexual Patterns" chapter into two groups of equal parts, one with data from men older than thirty-three years of age and the other with data from men thirty-two and younger. The former group had a median age of 43.1 years (and were most sexually active between 1910 and 1925) and the latter 21.2 years (and peaked sexually sometime after 1930). Kinsey found that the two groups had nearly identical behavior patterns in terms of age. He then divided his interview data again, and studied the relationship between the class that adult men (age twenty-five or older) were in at the time of their interview and that of their parents. Kinsey discovered, according to his data, that men's class-related sexual patterns were established by the age of sixteen. Even though they could not know in their teens in which class they would move as adults, they behaved sexually as if they were already members of that class as adolescents. Kinsey states, "The sexual history of the individual accords with the pattern of the social group into which he ultimately moves, rather than with the pattern of the social group to which the parent belongs and in which the subject was placed when he lived in the parental home."[38] How teenage boys would acquire knowledge of their future social class and how to emulate fellow men's sexual behavior is unclear.

If men's patterns survived through their late teens, men maintained them regardless of what class they entered later. If they changed patterns at all, that change happened in adolescence. "It is as though the bigger the move which the boy makes between his parental class and the class toward which he aims, the more strict he is about lining up his sexual history with the pattern of the group into which he is going to move." Sexual behavior pattern change, Kinsey found, had to take place in adolescence if it took place at all. Warner's and Lunt's findings on peer pressure among children may have explained such change. They describe how children of higher classes encouraged their lower-class peers to behave like them. Kinsey does not cite Warner and Lunt in his conjectures about how children might learn and mimic the sexual behavior of the class that they would move into as adults. However, considering that peer pressure was an instrument for shaping social values and behaviors along multiple axes (and at multiple ages), it was unsurprising for Kinsey to find that children in the class with more social power influenced the sexual development of their peers and future adult peers. "Children are, on the whole, conformists," he corroborates.[39]

While Kinsey's assertions about class movement were bold, he often based them on less than one hundred cases. For example, he states that "by early adolescence, the boys from class 4 [semi-skilled labor] homes who are destined to reach class 7 [upper white collar] may already be identified by their high frequencies of masturbation and by their very low frequencies of intercourse." There were seventy-one men who moved from class four to class seven by age twenty-five in the example above.[40] Kinsey goes into extensive detail about the number of individual cases he needed for each of his ultimate groups, but he did not have a set number of cases established to support his other generalizations. After the *Male* volume was published, an American Statistical Association review team would focus on the instability in the numbers of cases Kinsey used to make a generalization about certain groups.

Kinsey argues for the stability of class patterns across time, even with smaller numbers in each of the class samples than he may have wanted. He questions why other scholars would think that World War II would have a major impact on men's sexual patterns, when his evidence shows that it did not—men in wartime had the same types of experiences at the same ages that they would have had at home, just in different places and cultural contexts. Further, laws and Progressive Era social reform programs had had little, if any, impact on the sexual behavior of his interviewees. For Kinsey, as for Sorokin, the sexual values attached to each class were stable. Education was the element that facilitated an individual's movement between classes, not social reform or social upheaval. Kinsey speculates that the GI Bill would augment the number of college-educated men in the United States, and that there might potentially be broad social change driven by the presence of an increased number of college-educated men. But for Kinsey, classes and class differences as they manifested in sexual behavior would not change.[41]

There are more problems with Kinsey's class analysis than the small number of cases that he used to make broad statements about class mobility. Studying the parental home as one class lacked nuance. Kinsey did not consider that young boys might respond to the effects of class differently if parents had different class backgrounds. Yet another difficulty is that Kinsey made his interpretation of social stability of sexual patterns only through data from single men. He does not address the question of what happened to men who married women of different classes and how the man's (or the woman's) behavior changed in terms of class. Lastly, most of his class mobility data stops at age twenty-five, so there is little support for his assumption that for most men class status was fixed at age twenty-five.[42]

Kinsey stresses throughout the chapters on class the importance of gossip, the peer group, and social control in terms of keeping the behavior of adolescents and adults congruent with those of their peer groups. Peer pressure is also important in preserving religious restrictions. Kinsey almost seems disappointed in his findings regarding how his subjects adhered so strictly to class mores; that they were unwilling to try behaviors that their peers did not do. Of course, Kinsey found some variation in every group that he studied, but his data show that few men strayed far from the norms of their class. Class values structured sex education as well, whether the child acquired that education from parents, peers, religious instructors, or parents. Sex education and the class mores transmitted through them were "powerful enough to force most children . . . into becoming conforming machines which rarely fail to perpetuate the mores of the community." That was true also of adults, as "each adult lives and moves and does his thinking, to a considerable degree, in accord with the movements and the thinking of other persons who have about the same education and who usually belong to the same occupational class."[43] Human beings do not behave as machines, programmed to act without self-determined, independent thought, but machine-produced data showed the peer and social pressures on individual boys and men to act just like other members of their social class or face potentially harmful consequences.

In the *Male* volume's chapter on total sexual outlet, Kinsey muses about how harmful classifications of behavior often change lives for the worst. "Scientific classifications," he points out, "have been nearly identical with theologic classifications and with the moral pronouncements of the English common law of the fifteenth century."[44] He challenges his readers, and specifically his scientifically educated readers, to examine the reasons why they ceded understanding of and scholarship on sexual behavior to nonscientists. Scientists, he believed, have the right, if not the obligation, to conduct research on any aspect of the world's functioning. Scientists operate within structures of cultural and political power, Kinsey knew, and their research in this arena could change the classifications of sexual behavior to afford less harm to those with "deviant" sexualities. About a third of the way through the *Male* volume, then, Kinsey sets up the reader for the chapter on homosexuality and the 0–6 scale near the end of the book, the alternative classification of sexual behavior that he made a centerpiece of his thinking.

Kinsey's discovery of class conflict over sexual behavior, and his understanding of the ways that classification could help or could harm people, may have precipitated the statistical tables at the very end of the *Male* volume. One reason Kinsey included such detailed tables was to aid clinicians. Knowledge of group mores could help a clinician redirect a person's behavior to match the behavior of the social group, not to match that of the clinician, who was usually of a higher social level. Kinsey states, "Many clinicians feel that any re-direction of behavior should be limited to fitting the individual into the pattern of the particular group to which he belongs."[45] The tables contain averages of behaviors for specific groups according to the data that interviewees with those same characteristics had provided. The tables are broken down by all of the variables whose effects on behavior Kinsey describes in the main text. For example, if a twenty-four-year-old man was single, Jewish, devout, college-educated, and rated a "one" on the 0–6 scale, he could find the average amount of the nine types of sexual behavior for men with those characteristics.

The tables may have provided reassurance for men who wondered if their sexual behaviors were "normal" for their peer group. As Kinsey wrote in the introduction to the tables, "many persons who are disturbed over items in their sexual histories may be put at ease when they learn what the patterns of the rest of the population are, and when they realize that their own behavior has not departed fundamentally from the behavior of most persons in their social group."[46] The tables could have helped readers feel less alone, but they also may have made them feel more isolated if their behaviors were not in the normal range. Kinsey generally encourages readers not to feel artificially restricted by the broad social mores that clash with their personal desires, such as desires for premarital intercourse. But he suggests in the introduction to the fifty-six-page chapter on the statistical tables that adherence to class mores would minimize class conflict. Kinsey valued individual variation in sexual behavior, but only if it did not cause harm. If a man's departure from class patterns led to unhappiness or pain, then Kinsey thought he should consider adhering to the mores of his class. "Personality conflicts more often depend upon the individual's departure from the pattern of the social group to which he belongs, less often upon his failure to conform to the publicly pretended social code or to the formulated laws," he wrote.[47] Kinsey saw that conformance to class norms—or consciously becoming a "conforming machine"—was the best way to avoid conflict in one's sex life and social life.

The data Kinsey had collected on class and educational level reveals the strong effect that those social factors had on men's sexual behavior patterns. Though Kinsey continued to assert that each individual's behavior is unique, a man's social class shapes his decisions about how to behave sexually. Class differences also influence how members of some classes perceive and police others. Peter Hegarty has written that "Kinsey's class-inclusive sample told a story—not of intergenerational stability—but of uneducated men who were dissolving the norm that legitimated sex through the bonds of matrimony."[48] The small numbers of men who moved into different social classes as they became adults indicated to Kinsey that most men adopted the behavior patterns of their peers, and that they preserved class patterns as they aged. Kinsey included statistical tables of behavioral changes for the benefit of clinicians and those interested in seeing if their behaviors were "normal" for their peer group. Kinsey's interpretations of his social class data were his attempt to combine his data, statistical method, and background reading on education and class in order to propose a theory of how class standing affects sexual behavior and social life.

More broadly, Kinsey's work on social class indicated that tracing how behavioral patterns are perpetuated over time and become behavioral standards were matters of scientific interest. As he wrote in the opening of the first chapter on social class, "These questions [about the impact of social class on sexual behavior] are of such social significance that it is high time that scientific data replace the loose statements and easy conclusions drawn by persons who find some sort of advantage in bewailing the ways of the world."[49] If class-based sexual behavior patterns were replicated across generations, Kinsey thought that could lead educators, physicians, policymakers, and anyone interested in the effects of sexual behavior on social life to rethink their approach to teaching and policing sexual behavior. As the book progresses, he shows how a revised classification of sexualities through qualitative scientific methods could alter ideas and ideals of behavior and identity.

Through Punched-Card Machines to the 0–6 Scale: Toward a New Normal

From time to time throughout the *Male* volume, Kinsey reminds his readers that classification is a skill and an art, rather than a given, and that the sex history interview data presented has been processed via punched-card machine. Kinsey had used a paper form for organizing data about galls and gall wasps, but he had not needed or wanted to use many further mechanisms for analysis. However, for his sex research data, the use of machines made providing mathematical correlations between behaviors and characteristics possible, even if not all readers agreed with his interpretations. The combination of mass data, classificatory expertise, and a set of machines showed that these elements together could bring new insights to legal, cultural, and political considerations of sexual behavior. The numbers that came from processing data on punched-card machines supported Kinsey's ideas on the relationship of orgasmic premarital sex to "marital adjustment," his repeated observations of how unscientific classification of people and their behavior usually harmed more than it helped, and his clearest response to damaging forms of sexual classification: the 0–6 scale. By reclassifying sexual behaviors, desires, and fantasies as normal across a spectrum of opposite-sex to same-sex, Kinsey imagined the ways that classifications of sexual behavior could be liberating rather than punitive.

The *Male* volume's chapter on social level and sexual outlet contains just one of many examples throughout the text of how Kinsey and the ISR staff's use of punched-card machines influenced their data organization, analysis, and interpretation. Kinsey and his coauthors put in table form statistics on heterosexual coital techniques and nudity in groups representing three educational levels: eighth-grade education or less, high school education, and at least some college or more. The data in the chart show that side-by-side and female-superior positions were the most frequently used positions across ages and educational levels aside from the so-called biologically natural male-superior position. Men with at least some college education in the adolescent through age twenty-five group were most likely to use a female-superior position and least likely of any age or educational group to engage in coitus in a standing position.[50] Kinsey and the staff could then use such data to hypothesize about the possible age- and education-related reasons for such a behavioral discrepancy. They made abundant use of the particular ability that the punched-card sorters had to place different elements of hundreds or thousands of sex histories alongside each other to suggest relationships between them. Punched-card machines made these calculations available with relatively little physical effort and hand calculation. Further, once the staff punched the cards for a history using the key punch or gang punch, they could easily change the brushes on the sorter to gather correlative data for another two- or three-element analysis. The manipulability of the sorting brushes made the creation of mass correlative data a relatively simple task to execute with thousands of histories. The machines made possible new kinds of research questions as well as the data to answer them.

Furthermore, Kinsey was able to use machine-created data to correlate rates of heterosexual petting, premarital intercourse, and divorce rate to disprove the notion that petting and premarital intercourse would lead to higher rates of divorce. In fact, he found the reverse—that individuals making premarital sexual "adjustments" and gaining some knowledge of sexual behavior before marrying had a smoother transition to married life and were less likely to divorce. He states that while he would save a full accounting of the role of sexual adjustment in successful marriage for a subsequent study, he could nonetheless conclude from his preliminary results that "sexual maladjustments contribute in perhaps three-quarters of the upper level marriages that end in separation or divorce."[51] Middle- and upper-class people were most likely to adhere to the sexual restraints their families and peer groups established and enforced in their premarital years and found it hardest to let them go after marriage. Machine-produced data made that conclusion, and hundreds more in the *Male* and *Female* volumes, possible. The inescapable presence of machine-produced data in the Kinsey Reports emphasizes the importance of quantitative methods in the human sciences.

In the book's first chapter, Kinsey claims the right as the scientist doing the data collection to interpret his data in a way that the data supported. Kinsey knew, of course, that readers, including other scientists, would disagree with him on any and every aspect of the book. He felt, however, that his deep knowledge of the data, its machine processing, and his skills in classification gave him the right to interpretation. In the beginning of the chapter on adolescence and sexual outlet, he returns to the idea that figuring out the relationship of personal characteristics and behaviors, let alone determining causal factors, is complicated and fraught with potential errors. Kinsey returns to his intellectual grounding in classical taxonomy to point out the difficulty of identifying the onset of adolescence with precision given the many factors that indicate a young person has reached adolescence. He uses the opportunity to address the broader problem with isolating one physical or physiological factor to determine a complicated change of state: "The history of systematic botany and systematic zoology is replete with attempts to discover significant and diagnostic characters which might provide clear-cut and absolute bases for systems of classification; but the modern taxonomist finds that the use of a single character inevitably provides a classificatory system which is artificial and, at least at certain points, in direct conflict with data from other sources."[52] To reduce the possibility of errors, scientists needed to understand their own approaches to examining data and to arguing for relationships among characteristics, and to be aware that the processes of lumping and splitting characteristics required a comprehension of the multiplicity of factors at work. Even then, the scientists' interpretation of data could be incomplete.

Kinsey knew well that correlation was not the same task as causation in the interpretation of data with numerous intersections and characteristics, even though some readers and critics accused him of doing just that. He uses a discussion of classification to argue that no interpretation of data is complete or perfect, including his own. "The use of multiple characters in a taxonomic classification inevitably calls for a certain exercise of subjective judgment . . . but the errors introduced by judgment are not likely to be as misleading as the artificialities introduced by the use of a single set of criteria in classification," he wrote.[53] He reminds his readers that they too may be using "a single set of criteria"—perhaps their personal experience or individual worldview—to judge the accuracy of his interpretations. His repeated reminders that classification of scientific data is difficult, and needs a first look by someone with trained judgment, were no accident. As the *Male* volume proceeds to cover the different types of sexual outlet, he deliberately pairs his ideas of the scientific classification of sexual behavior with historical and present-day moral and religious classifications of sexual behavior. Describing the detrimental effects of moral and religious classifications of behavior, and their effects on men's and women's health and well-being, set the stage for the radical reclassification of behavior evident in the 0–6 scale.

Men's and women's sexual problems, he found, were strongly based on how their social, legal, and religious systems taught them to classify their sexual feelings and behaviors. Those systems, using what was in Kinsey's view pseudo-scientific evidence and reasoning, taught that many sexual behaviors and desires that were common to humans across the world and to many mammalian species were sinful, evil, and illegal. Kinsey hints throughout the *Male* volume that such classifications are deeply damaging both to men and women who try to uphold those ideals and to those who defy them. At the end of the chapter on heterosexual petting, he describes how they have affected the everyday worldview of the average American man:

> As an educated youth he has acquired ideas concerning esthetic acceptability, about the scientific interpretations of actions as clean or hygienic, about techniques, mechanically, when he has intercourse. He has decided that there are sexual activities which are right and sexual activities which are wrong, or at least indecent—perhaps abnormal and perverted. Even though these things may not be consciously considered at the moment of intercourse, they are part of the subconscious which controls his performance. Few males achieve any real freedom in their sexual relations even with their wives. Few males realize how badly inhibited they are on these matters.[54]

Such moral training affected individual men and women as well as the institution of marriage. Kinsey's data suggested that premarital petting to orgasm prepared people for more orgasmic sex in marriage. Marriages could be strengthened and the divorce rate could decrease with improved scientific knowledge of the wide range of sexual behavior. Women were even more hampered in their expressions of sexual behavior and desire by contemporary classifications of "good" and "bad" sex. They had even less access to information than did men, and the information they had was often incomplete or inaccurate.

> There are numerous divorces which turn on the wife's refusal to accept some item in coital technique which may in actuality be commonplace behavior. The female who has lived for twenty or more years without learning that any ethically or socially decent male has ever touched a female breast, and the female who has no comprehension of the fact that sexual contacts may involve a great deal more than genital union, find it difficult to give up their ideas about the right and wrong of these matters and accept sexual relations with any abandon after marriage.[55]

Kinsey uses these descriptions of individuals to underscore how unscientific classifications of behaviors have harmed individuals, marriages, and society as a whole. If people learned instead that there were endless combinations of behavior, desire, and fantasy that existed in the human population, they would learn that their feelings and those of their willing partners were "normal." Only celibacy and delayed marriage, as Kinsey had argued in the IU marriage course ten years earlier, were unnatural forms of consensual sexual behavior.

After one chapter each on premarital, marital, and extramarital sexual behavior and sex with prostitutes, Kinsey then turns to an analysis of homosexual behavior, where for the first time he argues for placing all individuals somewhere on a 0–6 scale, a measuring instrument that he had developed nearly eight years earlier and refined over time. The 0–6 scale illustrates the ranges of heterosexual and homosexual behavior, desire, and fantasy in an individual's sexual history, and it had been percolating in Kinsey's mind for more than seven years before it was in print. Kinsey created the 0–6 scale with the inspiration of Katharine Davis's work on women's sexual behavior and the help of the IU graduate student Glenn Ramsey. The 0–6 scale may also have been Kinsey's rebuttal to the Terman-Miles masculinity–femininity scale, a popular diagnostic tool that graded individuals on stereotyped gendered characteristics. The scale considered "masculine" women and "feminine" men as "inverts" or homosexuals, and the scale's creators encouraged the diagnostician to help their patient subscribe to codified gender roles.[56] The idea for Kinsey's scale first came to him in summer 1940, as he and Ramsey studied Ramsey's own interview data on the sexual behavior of preadolescent and adolescent boys in Peoria, Illinois. Ramsey found that about a third of his young subjects had a combination of homosexual and heterosexual contacts. As Kinsey's sex history interviews with IU students, faculty, and contacts in Chicago showed combinations of same- and opposite-sex behavior as well, he realized that he needed a means of studying and representing the diversity of behavior. He wrote in a letter dated September 20, 1940, to Ramsey about a version of the 0–6 scale and enclosed a working diagram. "I now think we have something which is perhaps the most significant thing yet available under the development of the heterosexual–homosexual picture." He had resigned from the IU marriage course only ten days earlier, but he did not mention that to Ramsey at all; he had already moved on from the experience.[57]

Two months later, in November 1940, Kinsey submitted an article to the new *Journal of Clinical Endocrinology*. It criticized an article by Clifford A. Wright that linked androgen and estrogen levels to "normals" and "17 clinically diagnosed male homosexuals." Kinsey attacks Wright's definition of homosexuality and his idea that a distinct type of person who could be called "homosexual" existed at all: "More basic than any error brought out in the analysis of the above data is the assumption that homosexuality and heterosexuality are two mutually exclusive phenomena from fundamentally and, at least in some cases, inherently different types of individuals."[58] Furthermore, Kinsey objected to the idea that "normal" and "homosexual" were two different types of men, and to the idea that such groups of men could be delineated by a certain amount of hormones in their bloodstreams. Kinsey hypothesized from the data he had collected from the marriage course, from homosexual and heterosexual men and women in Chicago and in northern Indiana, and from Ramsey's data that it was more accurate to apply the term "homosexual" to behaviors than to persons. Thus Kinsey derived one of his signature ideas—treating homosexuality as part of a heterosexual–homosexual continuum, rather than as a discrete anatomical, physiological, or social category—primarily from his own gathered data, and from Ramsey's findings that boys had sexual relationships with each other and with girls.

Kinsey knew early in his sex research career that his newly rented punched-card machines had transformed the way he was able to think about and to manipulate his data, and that the 0–6 scale had potential to be groundbreaking in how he and others thought about sexual behavior. As he mentions in a February 1941 letter to Ramsey, only a few months after first obtaining the tabulating machines, "we have just completed our first set of punched cards which covers the heterosexual-homosexual formulae for the entire lifetime of the individual," clearly connecting the presence of the machines with the creation of the 0–6 scale.[59] After the scale was published seven years later, it became Kinsey's signature theoretical model of human sexual behavior. The data supporting it were among the first batches he transferred to punched cards and tested for accuracy. The numerical machine-produced data corroborated the numerical 0–6 scale model that he had created by hand. For Kinsey, machine technology reinforced the patterns of sexual behavior that he had conceived of without it.

Gradually the scale that Kinsey first proposed to Ramsey in 1940, and hinted at in "Criteria for a Hormonal Explanation of the Homosexual" in 1941, was adjusted to look as it does in the *Male* volume.[60] It became a horizontal scale with zero as the indicator for heterosexual behavior and desire alone and six as the indicator for homosexual behavior and desire alone, with one through five standing for degrees in between. The scale encompasses many gradations of possible sexual experience. In other words, "The 'normal' was neither the average nor the median, nor even the ideal, but should be redefined as the entire range of experience itself." The scale modeled Kinsey's perception of how behavior, not identity, should mark an individual's sexuality, without stigma attached to any behavior, as "The scale was designed to eliminate the distinction between heterosexual and homosexual identities."[61]

Kinsey states that "the rating [on the scale] which an individual receives has a dual basis. It takes into account his overt sexual experience and/or his psychosexual reactions. . . . The position of the individual on this scale is always based upon the relation of the heterosexual to the homosexual in his history, rather than upon the actual amount of overt experience or psychic reaction."[62]

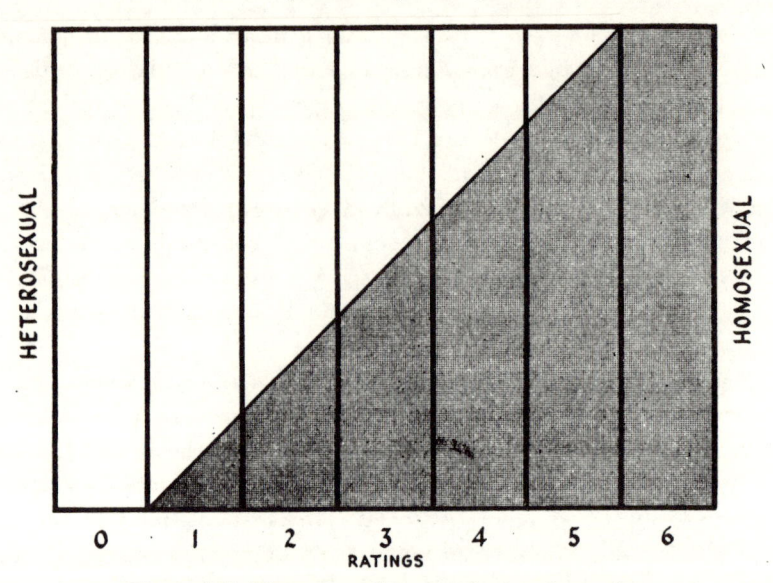

138 *The Taxonomy and Classification of Human Sexuality*

Figure 161. Heterosexual-homosexual rating scale

Based on both psychologic reactions and overt experience, individuals rate as follows:
- 0. Exclusively heterosexual with no homosexual
- 1. Predominantly heterosexual, only incidentally homosexual
- 2. Predominantly heterosexual, but more than incidentally homosexual
- 3. Equally heterosexual and homosexual
- 4. Predominantly homosexual, but more than incidentally heterosexual
- 5. Predominantly homosexual, but incidentally heterosexual
- 6. Exclusively homosexual

Figure 5.1. The 0–6 (heterosexuality–homosexuality) scale created from the Institute for Sex Research's punched-card data. Reprinted from Alfred C. Kinsey, Wardell B. Pomeroy, and Clyde E. Martin, *Sexual Behavior in the Human Male* (Philadelphia, 1948), 638. Reproduced by permission of the Kinsey Institute for Research in Sex, Gender, and Reproduction, Inc.

For most of Kinsey's subjects, the proportions of same- and opposite-sex desire and behavior were equivalent, but other times their behavioral history moved them toward one end of the scale and their history of desires moved them toward another. So Kinsey and his fellow interviewers determined their place on the scale when "one aspect may seem more significant than the other, and then some evaluation of the relative importance of the two can be made." "There are gradations in these matters," Kinsey wrote in the *Female* volume, so the scale needed to have multiple points to map them adequately.[63] Though the interviewer made the final determination of an individual's place on the scale for the purpose of the research, the interviewee's own perception of his or her place on it also affected the interviewer's decision.[64] Additionally, "Kinsey's was a universalizing epistemology of homosexuality that firmly rejected any ontological relationship between homosexuality and gender inversion."[65] Unlike one's place on the Terman-Miles masculinity–femininity scale, one's place on the 0–6 scale had nothing to do with upholding gendered or sexual stereotypes of behavior. Neither was any combination of sexual desire, behavior, or experience pathologized. Rather, the 0–6 scale was a level classificatory device that neither encouraged nor discouraged anyone from engaging in the behavior that they desired.

Kinsey argues that being a "one" on the scale means that the individual has "only incidental homosexual contacts which have involved physical or psychic response or incidental psychic responses without physical contact." Their homosexual experiences are accidental or unplanned, "inspired by curiosity," or "more or less forced upon them . . . when they are asleep or when they are drunk, or under some other peculiar circumstance."[66] Placement as a "two" on the scale means that a person has both opposite-sex and same-sex feelings and experiences, but that the former continues to outweigh the latter, and "if they have more than incidental homosexual experience, and/or if they respond rather definitely to sexual stimuli."[67] "Threes" have roughly equal same- and opposite-sex desires, showing no preference for one or the other, "and it is only a matter of circumstance that brings them into more frequent contact with one of the sexes."[68] A person who was a three could be married and have frequent intercourse with a spouse but have a strong interest in sex with one of their own gender, or be single and be sexually active with persons of the same sex yet intensely fantasize about and desire sex with a member of the opposite gender. Being a "four," "five," or "six" on the scale means being the opposite of a two, one, or zero, respectively. The amount of sexual desire or experiences does not matter in determining a person's place on the scale, only the proportions of desire and behavior relative to each other in the person's past and present.

Furthermore, given how people can move from one sociocultural context to another—and from one realm of sexual possibility to another—by moving to a new city, taking a new job, getting married or divorced, or making friends in a new social circle, one's place on the scale could, and usually does, change over a lifetime. The numbers zero to six are simplified placeholders for mapping the complexity of human sexual experience. As Kinsey puts it, "the reality is a continuum, with individuals in the population occupying not only the seven categories which are recognized here, but every gradation between each of the categories, as well."[69] Furthermore, Kinsey specifically objected to the term "bisexuality" to describe anyone between a one and a five on the scale, as "such a limited scale does not adequately describe the continuum which is the reality in nature."[70] For example, people could envision themselves partway between the two and three markers on the scale, but if the interviewer asked for a round number, they could provide one. Everyone could find a place on the scale, and no place had any sociocultural stigma or consequence, positive or negative. One could not be jailed, refused employment, or cast out from society simply for publicly being any number higher than zero, as many individuals with a single instance of homosexual behavior were in the 1930s through the 1950s.[71] Given that half of male interviewees had some same-sex sexual experiences and 28 percent of female interviewees did (37 percent and 13 percent to the point of orgasm, respectively), modeling sexuality on a numerical scale decoupled desire, behavior, and experience from identity and placed all behavior on an equal sociopolitical footing.[72]

The scale was meant to replace the tiered classification system that existed among men. Kinsey spends several pages articulating the current male sexual hierarchies in order to highlight the nonhierarchical clarity of his own 0–6 scale. He criticizes the idea that men could view mutual masturbation, oral sex, and anal sex between two men as heterosexual depending on the position of the men in the encounter. Men who held such views were deluding themselves, as their histories "show few if any cases of sexual relations between males which could be considered anything but homosexual."[73] Kinsey upended a classification system designed to uphold the sociocultural power of the penetrative man and to make the receiving man less powerful. The 0–6 scale dismantled a system that perpetuated inequalities between men who had sex with each other, and between those who had same-sex experiences and those who had none.

Kinsey's tabulating of interview data with a sorter had been congruent with his creation of the 0–6 rating scale. The correlation between these developments in Kinsey's work and thinking suggests that the enumeration of interview data into punched cards was a critical inspiration for the scale. The best-known statistic Kinsey obtained from the sex history interviews and the 0–6 scale became 10 percent: 10 percent of males interviewed were more or less exclusively homosexual for three years (rated a five or a six) between the ages of sixteen and fifty-five. The scale and "10 percent," both derived from punched-card machine data, would become two of the most visible and lasting elements of his analysis of homosexuality and of human sexuality in general.[74]

The *Male* volume highlights the interdisciplinary sources that Kinsey brought to his study of sexual behavior. Those sources provided him with the theoretical justification for studying natural phenomena like sexual behavior with quantitative methods. The statistical sex research sources structured how he solicited histories, from whom he solicited histories, how he interviewed individuals, how he tried to ensure truth-telling, and the questions asked in the interview. Biometric and statistical sources showed him how he could use mathematics to look for patterns among his data, and provided methods for him to order complex data and present the variation he found in a fashion that readers could understand and then use in their own work or lives.

Kinsey's interest in the effects of anatomy and physiology on sexual behavior would draw him into more complex research in a wider range of disciplines. After completing the *Male* volume, he still did not have a complete picture of how humans behaved sexually. The *Female* volume would be an opportunity to put together an even more comprehensive bibliography and synthesis of human sexual behavior, as it would focus on women's sexual behaviors specifically and also on comparisons of men's and women's behaviors. As he states in the social level chapter in the *Male* volume, "In terms of academic disciplines, there are biologic, psychologic, and sociologic factors involved; but all of these operate simultaneously, and the end product is a single, unified phenomenon."[75] The *Female* volume was another attempt to reach that "single, unified phenomenon."

6

The Boundaries of Sexual Categorization

The present volume constitutes the second progress report from the study of human sexual behavior which we have had under way here at Indiana University for some fifteen years. It has been a fact-finding survey in which an attempt has been made to discover what people do sexually, what factors may account for their patterns of sexual behavior, how their sexual experiences have affected their lives, and what social significance there may be in each type of behavior.

—Alfred C. Kinsey et al., *Sexual Behavior in the Human Female*

SEXUAL BEHAVIOR IN THE HUMAN MALE, released to the public the first week of January 1948, immediately generated numerous responses from readers across the United States and the world. The twenty-five thousand articles gathered by the newspaper clipping service that the Institute for Sex Research employed eventually filled seventy-two oversized binders.[1] The *Male* volume sold 185,000 copies in the first two months of publication.[2] Academic criticism of and interest in the *Male* volume was similarly extensive and swift across numerous academic fields, including a special conference organized by the American Social Hygiene Association devoted specifically to the volume (which Kinsey did not attend).[3] Kinsey became an instantly recognizable American household name; he received thousands of letters, and suddenly all kinds of people wanted access to him—offering their own sex histories or records of activity, wanting to confess their sexual indiscretions, asking for an interview to gain deeper insight into his research, or seeking counseling for personal sexual problems. Meanwhile, he and the ISR team kept taking histories of men and women, reading new research, and collecting art, books, and ephemera. They also began to research and write the *Female* volume, the second of the nine volumes that Kinsey had envisioned from the outset of his sex research project.[4] The research team knew that they would break new ground with another large study, now with 5,940 of the women's histories that they had gathered through January 1, 1950. Pomeroy remembered, "As we began to correlate the data for the *Female* volume, we experienced the exciting feelings of discovery in a relatively unexplored field that had occurred as we prepared the first volume."[5] *Sexual Behavior in the Human Female* was published on August 20, 1953, with great fanfare. Some writers dubbed it "K-Day," comparing the book's release to dropping an atomic bomb.[6]

The *Female* volume aimed to address broad questions regarding human sexual behavior: "discover[ing] what people do sexually, what factors may account for their patterns of sexual behavior, how their sexual experiences have affected their lives, and what social significance there may be in each type of behavior." That was a more far-reaching set of goals than those of the *Male* volume, which lists only the first of those two goals in its opening paragraph.[7] From the *Female* volume's first paragraph onward, readers knew that it addressed a more extensive set of issues, including not only what people did sexually and why but also how those experiences influenced their lives as well as the wider social impact and meaning of those experiences. Each of the volume's chapters on sexual behavior has in its second part a long summary comparing the statistics of the *Male* volume to those just discussed from women. Those lengthy side-by-side comparisons encourage readers to look for patterns and discrepancies in the interview data. Though the book is called *Sexual Behavior in the Human Female*, it contains analysis of sexual behaviors and bodies across the human race.

The *Female* volume, in its three parts: "History and Method" (three chapters), "Types of Sexual Activity among Females" (ten chapters), and "Comparisons of Female and Male" (five chapters), highlights Kinsey's ability to integrate material from an even larger range of sources than he had used for the *Male* volume, including not only his interview data but also numerous other resources and texts: over one thousand secondary sources in English, Spanish, French, German, Dutch, Latin, and Japanese. Kinsey and his increasing number of professional staff members and advisers more than doubled their sources from the *Male* volume to the *Female* volume, from 597 to 1,217 items. A larger set of sources and more interview data necessitated a more comprehensive and sophisticated approach to classification than the *Male* volume had needed. "Even though it has been difficult to quantify and statistically treat most of this recorded material," he notes in the *Female* volume, "it has been invaluable in its portrayal of the attitudes of the subjects in the study."[8]

Bearing in mind that complexity, this chapter addresses three different areas of Kinsey's classification practices in the *Female* volume, focusing on the various classifications that he made regarding men's and women's sexualities and men and women in general.[9] The book moves from considering similarity and difference between men's and women's sexual behavior to considering more radical ideas of similarity in their anatomy and physiology, including orgasm and gendered capacity for it. By the end of the book—when readers might have expected Kinsey to summarize his ideas of male and female sexual difference—he concludes that there may not be much difference between genders after all.

First, this chapter examines how Kinsey structured his description of differences in men's and women's sexual behavior throughout the central body of the text. In the ten "Types of Sexual Activity among Females" chapters—preadolescent sexual development, masturbation, nocturnal sex dreams, premarital petting, premarital coitus, marital coitus, extramarital coitus, homosexual responses and contacts, animal contacts, and total sexual outlet—Kinsey presents men's and women's similarities and differences by repeatedly comparing and contrasting male and female data. Second, Kinsey had read extensively in animal behavior and cultural anthropology when he wrote the *Female* volume. As a result, the *Female* volume contains many comparisons of the sexual behavior of white American women with female animals and women in non-Western cultures around the world, making "female" a broad category indeed. Thirdly, the last section of the book, with its five chapters describing the sexual human body from five different perspectives, examines bodies to look for conclusive markers of male and female difference. The classification of humans into two gendered groups that structures the previous thirteen chapters of the book becomes instead discursive fragments rather than physical realities.

Those different sets of classifications serve three purposes in the structure and organization of the *Female* volume. First, keeping comparisons between men and women ever-present makes it clear that the book addresses human sexual behavior generally, and that it does not just serve as a female mirror image to the *Male* volume. Second, Kinsey chose to lump the behaviors of white American women, non-Western women, and female animals together to mark the similarities between human sexual behavior and animal behavior, and to emphasize that many behaviors that were considered outside the realm of possibility for Westerners are part of female mammals' and many women's normal activities. Doing so made the category "female" more complicated, and drew out many possibilities for how women could think about their sexualities. Third, Kinsey found no clear way to split men and women into two groups using any one marker, be it orgasmic capacity, behavior, identity, or body composition. Analyzing the third section of the book shows how previously solid forms of classification in an author's thinking, such as a binary division between men and women, break down under the weight of contrary evidence, making room for new ways of thinking about how to classify human beings sexually. By the end of the book, Kinsey finds that divisions of human sexual difference other than gendered ones make the most sense.

Comparing Similarity and Difference

Issues regarding how to classify data by and about women, including their sex histories, diaries, artwork, advice literature, theological tracts, sex calendars, films, and other sources began well before Kinsey began writing the book. Kinsey, Pomeroy, Martin, and their new coauthor Paul Gebhard (who joined the team in 1947) had to decide which of the sex histories to focus on and which ones to save for later volumes. They made some decisions about what data to use that they do not clearly explain in the *Female* volume or elsewhere. The lead author's interests had clearly shifted away from the sex histories primarily and more toward the sex histories as they combined with other sources. However, choices about data to include, data to exclude, and which forms of analysis to use were not up to Kinsey and his research group alone.

The overall structure of the *Female* volume shows the impact of the three-man American Statistical Association team who reviewed Kinsey's statistical methods and findings at the request of the Rockefeller Foundation (Kinsey's primary source of funding) over a two-and-a-half-year period from 1950 through 1952.[10] The statisticians had no major objections to the data on homosexual behavior or to the 0–6 scale, and thus the scale and attention to homosexuality remained unchanged from the first volume to the second. And although Kinsey refused to write out the interview questions in full as they asked, he did write an extended account of the many sources of data in the *Female* volume's third chapter. Kinsey eliminated the comparative US Census data from the *Female* volume (the interview data that the researchers adjusted in the *Male* volume to reflect the US population proportionately) because they did not reflect the sexual behavior of the whole US population as accurately as either Kinsey or the statisticians wished.

Thus the role of statistics in structuring the two volumes' text changed between the *Male* and *Female* volumes. Kinsey's extended discussion of his statistical methods illustrates how much that chapter was his method of answering the ASA's criticisms. While Kinsey and his team continued to take accurate statistics seriously, the chapters using interview data—unlike those in the *Male* volume—were preludes to exploring bigger questions of sexual differences and similarities between men and women.

In the statistical chapter, Kinsey first outlines the parameters of probability sampling and then reiterates his reasons for not using it to gather the sample: the high refusal rates that the interview team would have encountered if they had tried to obtain interviews from random individuals instead of from groups with whom they had built rapport. A sex surveyor was not alone in facing high refusal rates for his questions. Economic, social, and political researchers also heard a lot of "nos" from potential interviewees, Kinsey wrote, mindful of an interdisciplinary audience. "Our experience leads us to predict that the attempt to secure sex histories from lone individuals would have resulted in refusal rates so high that the sample would have been quite worthless," Kinsey argues. As accurate reporting and accurate sampling were "to some extent antagonistic," the team used 100 percent sampling as a compromise method. Kinsey also made some smaller changes per the ASA team's request.[11] He was reluctant to alter the statistical methods that had been in place for a decade. Where he and the statisticians disagreed, Kinsey dropped most of the sections (social level and stability of sexual patterns) that caused them concern.[12]

One of the most obvious differences between the two Kinsey Reports, aside from the addition of the five chapters on "Comparison of Female and Male," is the nearly complete disappearance of social class in the *Female* volume. Where the *Male* volume has one chapter each on social level and the stability of sexual patterns across classes, the *Female* volume contains minimal coverage of the impact of social class on women's sexual behavior. There were several reasons for that shift in treatment. First, the ASA team and the other statisticians that Kinsey and his team consulted had detailed concerns regarding the *Male* volume's chapter on stability of sexual patterns. Second, Kinsey's team had not found that class level had an impact on women's sexual behavior, and they were not very interested in explaining the lack of a particular influence instead of the reasons for an influence. In addition, women's class status, they thought, was more difficult to determine than men's social class. They felt that women's class status had a greater variety of sources than did men's class status, and that it was more volatile and likely to change over time because of marriage than because of changes in employment or education level. To determine women's social level required factoring in the class of her parental home, and, if married, the class into which she married.[13] Her self-perception of her own class was also relevant. Of course, it seems unlikely that class had no impact on women's sexual behavior given the strong impact it had on men's sexual behavior, though Kinsey repeats the lack of class impact on women in each of the *Female* volume's chapters on behavior.

In part, this insistence on omitting class may have had something to do with subject sampling methods. Kinsey also made the choice to leave the prison sample out of the *Female* volume's calculations. The female prison and nonprison samples, Kinsey wrote to ASA review team leader William I. Cochran, "show great differences" and muddled the results of his calculations from the better-represented high school– and college-educated groups.[14] So the *Female* volume contains a more homogenous sample than the *Male* volume does; among other things, it excludes the 915 incarcerated women interviewed, as "their inclusion in the present volume seriously would have seriously distorted the calculations on the total sample." Like the *Male* volume, the *Female* volume does not include sex histories from African American or other nonwhite women.

The stated reason for the omission of data from nonwhite women is that the sample (934 cases) "is not large enough to warrant comparisons of the subgroups in it."[15] Although there is no documentation on this point, Kinsey's decision may have been a way for him to circumvent potential criticism that the volume's findings of significant amounts of nonmarital, noncoital behavior data were from prostitutes, criminals, or African American women, given broad stereotypes in white American culture about African American promiscuity and white female purity.[16] Kinsey challenges white female purity myths throughout the book by pointing out that many of the white women whom he and the team interviewed participated in "covert" sexual culture by having great amounts of many kinds of sex while appearing to uphold the restricted sexual norms of overt American culture.[17]

So some of the book's most sensational findings, such as that 40 percent of interviewees had premarital coitus to orgasm, 13 percent had same-sex experience to orgasm, and 26 percent had extramarital coitus to orgasm by age 40, could not be attributed to prostitutes or women convicted of crimes.[18] Kinsey was not entirely happy that more of his data could not appear in the *Female* volume, or that it would have to wait for one of the future volumes in his planned multivolume series.[19] He concludes in a letter to Cochran about the ASA team's early suggestions about the female sample: "We regret, of course, that the female volume will have to go out in this form but we think there is good reason why we should present the material that seems adequate and accept the fact that we cannot tell the whole story at this time."[20] Kinsey does not say specifically that a more thorough analysis of women's sexual histories—using all of the available data—would be even more startling to readers and more subject to politically motivated attack than the white, nonprison, mostly high school– and college-educated, and middle-class women's data on which he and his team focused. Thus the authors knew that the data in the *Female* volume was not "the whole story," but it was the story they were willing to discuss politically.

The authors found in their histories that date of birth, rather than social class, was the strongest outside influence on women's sexual behavior.[21] That discovery, however, did not lead to the creation of a chapter devoted solely to generational change in women's sexual behavior. Perhaps Kinsey and the research team did not want to take the time to investigate the more complex ways that class and date of birth affected women's sexual behavior. It is more likely that Kinsey and the research team were simply more interested in the third part of the *Female* volume, "Comparisons of Female and Male," than they were in probing deeper into their sex history data on women. The evidence for that shift in interest is reflected in a July 1952 letter that Kinsey wrote to Karl S. Lashley, his friend Frank Beach's doctoral dissertation adviser and a member of the Committee for Research in Problems of Sex's medical division. Lashley visited Bloomington twice, in September 1952 and March 1953, at the request of the publishing house W. B. Saunders, to give Kinsey and the research team his advice on the psychology, neurology, and endocrinology chapters.[22] In the letter's postscript, Kinsey reflects, "This aspect of our research seems to us of tremendous importance. In fact I am inclined to believe that it is more important than the population survey which is part of our work, and we have a tremendous body of previously unavailable data on this aspect of the physiology."[23] As gathering, analyzing, and publishing new data about the natural world were hallmarks of Kinsey's whole career, it is not surprising that Kinsey focused greater attention on his newer anatomical and physiological data instead of on his older sex history data. The *Female* volume was a mixture of both, as he needed both to address the problem of sexual difference between men and women.

The idea that gendered sexual difference had some basis in psychology first appears in the *Female* volume's discussion of men's and women's sexual fantasies. In the chapters on masturbation and nocturnal sex dreams, Kinsey addresses the varying role of fantasy in men's and women's sex lives. Masturbation and nocturnal sex dreams together, as non-partner-dependent acts, were "the activities which provide the best measure of a female's intrinsic sexuality."[24] Kinsey rarely uses the word "sexuality" in a book devoted to sexual behavior specifically, but its use in the chapter on nocturnal sex dreams shows his intellectual struggle with defining and capturing the full meaning of the broader term. Women had fewer nocturnal sex dreams than men overall, but it was the only behavior in which variation among men was greater than the variation in women.

While more men had nocturnal sex dreams and fantasized while awake—highly educated men most of all—only women could orgasm with fantasy but without their own or another's touch. Kinsey notes such a difference in order to make a point about differences in men's and women's sex-related psychology. Kinsey argues, with little evidence beyond his own opinion, that even though women have an imaginative capability that no man could match, men's ability to have a greater amount of nocturnal sex dreams meant that psychology influenced their sexual lives more than their anatomy, physiology or hormonal balance. For Kinsey, women's ability to fantasize to orgasm without touch while awake was less of a sign of imaginative capability than men's fantasizing to orgasm with touch. In other words, Kinsey uses those examples to illustrate differences in men's and women's psychology: while men need outside stimulation (visual, with their hands, or another object) to condition their imaginations so that they can fantasize to orgasm, women's ability to fantasize to orgasm is only a matter of properly focusing their minds.

Nonetheless, the ten behavior chapters contain Kinsey's thinking on other factors that shaped and affected women's sexual behavior. Those chapters also show that Kinsey had decided that male and female sexual differences were based on their psychology. That assumption led Kinsey to explain male and female sexual differences in the behavior chapters through a psychological lens, even if other explanations fit the data better. In the chapters on masturbation, nocturnal sex dreams, petting, and premarital coitus, Kinsey uses interviewee data to convince the reader that male and female sexual differences are psychologically based. The argument he makes in the sixteenth chapter of the book, on sexual psychology, is incorporated into the earlier chapters piece by piece, so the reader will not be surprised to encounter it later.

Kinsey did want to highlight that he had plenty of evidence that men and women are similar anatomically and physiologically, and that evidence was clearest when comparing men's and women's experiences of masturbation. Masturbation was "the most clearly interpretable data which we have on the anatomy and the physiology of the female's sexual response and orgasm."[25] Masturbation was a better indicator of female interest in orgasmic sex than intercourse was, as the latter usually required a partner's interest and focus on the clitoris to attain. Men began masturbating at adolescence, and women began doing so when they learned it was possible—sometimes from other women, but most often by exploring their own bodies. Women had orgasms as quickly as men did when masturbating, though they did not orgasm as much in coitus "due to the ineffectiveness of the usual coital techniques" for them.[26] While that evidence supports a concept of men's sexual behavior as hormone-driven and women's as more driven by the excitement of breaking social taboos against women's self-pleasuring, Kinsey sidelined such an interpretation in order to focus on how the physiological processes of masturbation were an indicator of physiological and gender equality.

Psychological explanations of male and female sexual difference made the most sense to Kinsey in light of his understanding of animal sexual behavior. Dependence on that explanation, though, led him to misinterpret and to misread data about women's sexual behavior. He shoehorned psychological difference into discussions not only of masturbation, fantasy, and nocturnal sex dreams but also of premarital petting and other partnered behavior. For example, in the premarital petting chapter, he argues that men, "and particularly American males, may find considerable psychologic stimulation in touching and manipulating the female breast," but why breast stimulation would be a strictly psychological phenomenon is unclear. Later in the same chapter, he again posits that the gender differences in arousal from oral-genital contact "may depend primarily on the greater capacity of the male to be stimulated psychologically."[27] Other explanations could better explain gendered sexual phenomena, though they did not fit a psychological framework. For Kinsey, the clarity of an explanation based on animal evidence trumped explanations that could have included more systemic sociocultural analysis.

Lastly, psychology was a discipline that depended on bodily manifestations of invisible brain and nervous functions. Perhaps it made the most sense for Kinsey to base his case for the gendered difference that he could not explain on psychological processes that he could not see and thus characteristically did not wholly trust. Or else overall gender distinctions, and "maleness" and "femaleness" as characteristics attached to specific persons, were making less sense to him as his research project moved forward.

After becoming close colleagues with the animal behaviorist Frank Beach, Kinsey's interests had centered less on the statistical interpretations of his research project and more on the bodily processes involved in sexual behavior.[28] As Kinsey states in the *Female* volume's statistical chapter about the last section of the book, "While we might have secured better statistical data on the incidences of the various types of sexual activity, we would not have arrived at our present understanding of the other, non-statistical problems which we have considered in equal importance to the present study."[29] Kinsey's incorporation of animal and anthropological material into the chapters on behavior set the stage for the third part of the book and a more expansive analysis of gendered sexual similarity and difference.

Animals, Americans, and Anthropological Studies

Kinsey's interest in finding the reasons for male and female sexual similarity and difference in bodies, minds, and social lives is a key narrative throughout the *Female* volume. Kinsey pays close attention in the *Female* volume to identifying patterns in his sample of nonprison white American women that match data on female animals and studies of women in non-Western cultures. He uses animal and anthropological data to suggest that American society might have less tension between its overt and covert sexual cultures if American sexual mores followed cultures with more relaxed sexual attitudes, and if they matched Americans' true animal natures—up to a point. Using patterns in animal sexual behavior to support patterns in human sexual behavior left Kinsey open to the charge that he was "animalizing" white Americans the way critics accused him of "insecticizing" them in the *Male* volume. For Kinsey, expanding the categories of "male" and "female" when studying human sexual behavior meant including an analysis of all types of humans and their mammalian cousins in the animal kingdom. Such a broadening of gender categories would ideally show how and why female and male animals were alike and different sexually.

Kinsey had begun reading anthropological and animal-related sex behavior books, as well as collecting films and corresponding with researchers in those fields, toward the end of the *Male* volume's preparation in 1947. Kinsey's goal in gathering sex-related research in anthropological and animal-related fields was to place white American sexual behavior in a comparative context. Including a discussion of sexual practices common in non-Western and non-American cultures would demonstrate the contingency of everyday American mores and would suggest a general lack of knowledge regarding standard mammalian behavior. Knowing what was common among mammals could reduce stigmas attached to common human behaviors, and knowing what was common for non-Western groups could inform readers how other cultures organized their sexual lives, often with less discord than Americans did.

Kinsey found evidence of all the behaviors he studied in Americans in non-human mammals. He additionally found them in what he termed "primitive groups" or "pre-literate people," principally natives of the Pacific Islands and central Eurasia and American Indians. Whenever he uses anthropological data in the *Female* volume, he points out its fragmentary nature, due to investigators' reticence to observe or to ask their human subjects about their sexual practices. The obvious difference between human and animal studies was that humans marry and animals do not, and so the concepts "premarital" and "extramarital" did not apply to animals. Scientists could, however, observe how animals teased or played with each other before coitus and the extent to which they mated repeatedly with the same partner. Similarly, animals were not able to recognize when their behavior was "homosexual" or "heterosexual," as researchers had labeled same- and opposite-sex behaviors.[30] But scientists could monitor when and how animals attempted to mate with members of their own sex. The eight chapters in part two of the *Female* volume that cover specific behaviors—masturbation, nocturnal sex dreams, premarital petting, premarital coitus, marital coitus, extramarital coitus, homosexual responses and contacts, and total sexual outlet—have some discussion of mammalian and non-Western cultural comparisons.[31] Kinsey was one of the first scientists to bring together anthropological and animal sex data with original sex-related data from white Americans to produce a detailed and complex picture of human sexual behavior.

Kinsey found evidence of female masturbation in many animal species and human cultures. He discovered that female elephants, horses, porcupines, skunks, guinea pigs, and monkeys all masturbated. Albert Shadle, a biology professor at the State University of New York in Buffalo, was especially helpful for providing—and discussing his observations of—sex films of porcupines, skunks, and raccoons that he had caged in his laboratory. When Kinsey had viewed animal sexual behavior in person or on film, or if another scientist confirmed his observations of the behavior, he usually mentions it in the *Female* volume's footnotes as well.[32] He found even more ample evidence of masturbation in women than he did in female animals.[33]

Kinsey's research into the nonmarital sexual behaviors (including masturbation) of non-Western women led him to consider the relationship of power to the monitoring and control of sexual behavior. Because most non-Western patriarchal cultures did not consider female masturbation threatening to their social order, it was not condemned, but other kinds of behavior were more heavily policed. "Human males throughout history and among all peoples have been most often concerned with the sexual activities of the female when those activities served the male's own purposes," he points out. Anthropologists had also observed Siberian, American Indian, and Pacific Islander women masturbating, but the true percentage of women who engaged in it was unknown, as the written record was "notably fragmentary and probably gives no idea of the true spread of the phenomenon."[34] The rate of masturbation that he found for the women in his sample was 62 percent, while men's rate was 93 percent. In Katharine Bement Davis's sample of unmarried women in the 1920s, 88 percent of those who masturbated reached orgasm. By juxtaposing masturbation by animals and non-Westerners with masturbation by white American women, Kinsey intended to highlight what he considered to be the artificial constraints on American girls and women, not to argue that animals and non-Westerners were alike.

Such constraints are also evident in the evidence Kinsey provides regarding premarital necking or petting, which he defines as "physical contacts between females and males which do not involve a union of the genitalia of the two sexes."[35] In the *Male* volume, Kinsey criticizes the extreme lengths to which American parents, educators, and religious leaders went to prevent premarital petting and sex play among young people. For the *Female* volume he sought to show that, given petting's ubiquity in the animal kingdom and its necessity in preparing mammalian bodies for intromission and ejaculation, social restrictions alone were not able to control precoital sex play. Textual and filmic evidence from the animal kingdom showed that most animals engage in some form of "petting" before attempting intromission.[36] Animals, including humans, undergo physiological changes such as elevated pulse and blood pressure rates, faster breathing, and vaginal lubrication in females as a result of preparation for coitus. "The human animal behaves like the mammal which it is when it engages in petting before it begins coitus," he wrote. The near universality of foreplay in mammalian species not only demonstrated to Kinsey that human adolescents should be less restricted in their petting practices, it also sparked his interest in investigating exactly what happened to animals' bodies as they readied themselves for intercourse and how those changes were or were not sexed.[37]

Premarital petting served as a socializing mechanism for humans as much as it did a preparation for actual intercourse. As Americans were disinclined to lower the age parameters for marriage to match the onset of sexual desires in adolescence, Kinsey urged lessening social restrictions on premarital petting that would also make for easier sexual adjustments to coitus in marriage. Kinsey found a number of non-Western cultures that allowed children and adolescents to experiment with sex play and to mimic adults without fear of recrimination. Those examples proved to Kinsey that many other cultures managed young people's interest in sex without causing them emotional distress or forcing them to sneak around behind their elders' backs for fear of punishment.[38]

Another aspect of nonmarital sexual behavior that Kinsey needed to include to make the *Female* volume comprehensive was extramarital sex. Kinsey ascertained from his anthropological readings that non-Western cultures had distinct and complicated rules governing extramarital sex, given that nearly all cultures had some form of marriage. Clellan S. Ford and Beach found in a comparative survey of seventy-six cultures that 39 percent allowed some form of extramarital coitus, mainly for men. Ford had found in an earlier study that about one-third of societies allowed it. However, the meaning of "extramarital" varied significantly, as did the extent to which the male or female partner strayed and the degree to which society policed or punished the straying. In general, men were allowed to have sex outside the pair bond without much recrimination, but women were punished for the same. Some cultures allowed women to have extramarital coitus as long as they hid it from their mates and society leaders.[39] Anthropological evidence showed that strong bonds existed between pairs of men and women across cultures, but also that those bonds were violated more often, and with greater impunity, by men.

Extramarital coitus was one type of behavior that did not have a clear analogy in the animal kingdom (a point Kinsey makes repeatedly in the book), and he struggled to find the examples and the language that would link the absence of lifelong bonding in mammals to the conflict between some individuals' desires for a variety of partners and the social (and usually spousal) demands for fidelity in human marriage. Even if he found the "bases of problems" in this instance to be "in the living stuff" of animals' bodies, it would be problematic to argue for dramatically upending the institution of marriage.[40] In the *Female* volume, he wanted to find a way to counterbalance the observation of multipartner animal mating and his belief in the institution of marriage—scant evidence of animal behavior alone could not justify loosening traditional forms of nuptial bonds.

In lieu of data on extramarital coitus in animals, Kinsey attempted to extrapolate from what few studies existed on animal pair bonding. His three-page discussion of the mammalian origins of extramarital coitus in the book, unlike his discussion of the mammalian origins of homosexual responses, includes two citations to forty-year-old studies with examples of what primate researchers described as "the increased vigor in copulation and the reduction of foreplay which characterizes a new partner." He had not found much literature on animal pair bonding in his research, but what he had found, to his mind, explained human desire for partners outside of marriage. "It is evident," he wrote, "that interest in a variety of partners is of ancient standing in the mammalian stocks," even though his scarce citations belie such an assertion.[41] Kinsey's argument for extramarital coitus, among all his reasoning for partnered sexual behavior outside marriage, is the least supported with evidence from the animal kingdom. Kinsey advocates that societies should uphold heterosexual marriage but also permit extramarital sex for "strong-minded and determined individuals" without grave consequence.[42] Only individual couples and societies, not animals, could rearrange customs surrounding extramarital sex.

Kinsey's evidence for the regularity of same-sex behavior in mammals was much stronger than his evidence for mammals' lack of long-term pair bonding. He uncovered a significant amount of evidence from scientists who recorded same-sex behaviors in animals of many species. Kinsey's wide variety of sources on the differences between male and female animals mounting members of their own sex would likewise spark his interest in the varying tendencies of men and women to engage in same-sex acts. Other surveyors of animal behavior contributed to Kinsey's long list of the same-sex actions of mammals including female martens, female cats, male lions, male elephants, male hyenas, male mice, and male rabbits. Scientists observed some male mice and rats ejaculating when they came in contact with each other.[43] Kinsey notes that some animals, such as male lions, male baboons, and male porpoises, have "exclusive, though temporary homosexuality."[44] He saw "no exclusively homosexual patterns" in female mammals, though some females in heat were indiscriminate and occasionally mounted males or other females when males were not mounting them.[45] His film collection also included depictions of same-sex animal behavior, and Shadle further informed Kinsey in personal communications of the homosexual behavior he had observed in male porcupines, male raccoons, and cows.[46] He also found data from anthropologists who observed that most same-sex activities were tolerated in cultures around the world. Ford and Beach calculated that 64 percent of the societies that they studied considered homosexual activity acceptable for certain persons, and that there was generally less female than male same-sex activity.[47]

Records of homosexual behavior in animals and in non-Western societies supported Kinsey's view that homosexual behavior is an organic part of the human sexual makeup; it is possible for everyone and an actuality for many. There were fewer records of homosexual behavior to orgasm in women's sex histories, but the variation in their behavior surpassed men's in terms of frequency and combinations of different behavioral types. Kinsey's data showed that 13 percent of all the women interviewees had homosexual behavior to orgasm by age forty-five. Women had less homosexual behavior than men did in their histories, a point that the data from animals and non-Western persons confirmed. "No exclusively homosexual patterns have been reported for female mammals," Kinsey states in a footnote, and he mentions his own observations of homosexual behavior in female rats and guinea pigs in another footnote.[48] The greater the abundance of same-sex behavior among animals, the stronger Kinsey felt arguing for its normalcy in American women's (and men's) behavior as well.

Kinsey does not mention in the *Female* volume that he not only collected stag and other pornographic films but also made his own films of human sexual behavior in addition to animal films. Bill Dellenback, the staff photographer at the Kinsey Institute, made the films of individuals, homosexual couples, and heterosexual couples, including Kinsey and other members of the ISR staff, in Bloomington and New York City. No one outside of the ISR staff and participants knew that the "attic films," as they called them, existed until Wardell Pomeroy (also a participant in the films) described them in his 1972 biography of Kinsey.[49] Making films of human sexual behavior made sense to Kinsey, as he wanted to be able to study human behavior using the same media that animal behavior scientists used. Neither Kinsey nor the animal behaviorists whose films he observed, including those of porcupine and raccoon expert Albert Shadle, could exhibit their films publicly. But in the last five, explicitly comparative, chapters of the book, Kinsey's use of animal data takes a different turn. His use of anthropological data mostly disappears, as most anthropologists did not measure their subjects anatomically or physiologically. Instead of using animal data to make human sexual behavior appear to be in concert with that of mammalian sexual behavior generally, Kinsey uses it to map the anatomy and physiology of sexual response and orgasm, ultimately working toward an understanding of the ways that men's and women's sexual behaviors differed or were similar and why.

Categorizing Men and Women: New Perceptions of Similarity and Difference

The addition of five chapters on anatomy, physiology, psychology, neurology, and endocrinology shifted readers' attention toward "what factors may account for [men's and women's] patterns of sexual behavior."[50] Kinsey argues that men and women are alike anatomically and physiologically but are different psychologically. In keeping with his long-held view that orgasm is a suitable measure for male and female sexual behavior, he makes a case that men's and women's orgasms are physiologically similar, and that clitoral as opposed to vaginal orgasms are the physiological reality for the vast majority of women. While some difference in psychological makeup makes the most sense for gender difference, he admits by the end of the volume that it is inadequate. In the chapters on the effects of the brain and hormones on sexual behavior, discussion of male and female sexual difference fades as Kinsey contemplates whether there is much use for gendered categories at all.

The strongest influences on the last five chapters of the *Female* volume overall were the two animal behaviorists Frank Beach and Karl Lashley. Beach, who primarily studied rats, insisted that all aspects of animals' bodies needed to be taken into consideration when determining why they did or did not participate in certain behaviors. A myriad of internal forces and external stimuli control sexual behavior for both male and female animals. For both Lashley and Beach, "complex instinctive and learned behaviors required an equally complex and integrated set of neural mechanisms for their mediation."[51] Beach and Lashley were such strong influences on Kinsey in part because all three men shared an interest in studying behavior across the mammalian kingdom, and in part because they were experts in areas of life science in which he was self-taught.

Kinsey uses the work of Beach and others interested in the connections between anatomy and psychology in chapter fourteen of the *Female* volume, "Anatomy of Sexual Response and Orgasm." He aimed to prove that men and women are alike in terms of their physical responses to arousal. Beach describes in "Review of Physiological and Psychological Studies of Sexual Behavior in Mammals," an article that Kinsey references repeatedly, how different forms of contact, noncontact, genital, and nongenital stimulation took place when mammals mated.[52] Kinsey shows how all parts of the body could serve as receptors of stimuli—to one degree or another. Kinsey wrote, "Our own observations on non-coital sex play cover the baboon, chimpanzee, monkey, dog, chinchilla, hamster, rat, guinea pig, rabbit, porcupine, raccoon, cat, mare, cow, sheep, sow, and skunk."[53] Beach describes how animals stroked or bit each other's ears; Ford and Beach confirmed Kinsey's finding that breast stimulation rarely aroused females; and Kinsey cites Shadle to support his statement that "manipulation of the partner's body with the nose and lips has occurred among nearly all the species of animals which we have observed."[54] Male and female animals were found to have many of the same precoital rituals that humans do.

The *Female* volume's chapter on the anatomy of sexual response and orgasm illustrates the similarities of men's and women's bodies as sexual and particularly orgasmic entities. Sophia Kleegman, a gynecologist and a member of the American Association of Marriage Counselors, wrote to Kinsey in November 1950 to say, "For almost two years I have been wanting to write you a 'long letter' about female orgasm, hoping that because you not only have vast experience, but more especially because that experience is distilled through clear thinking, perhaps at long last, female orgasm could be interpreted on a basis of physiologic, realistic understanding." Kinsey asked six gynecologists, including Kleegman, to do research on vaginal sensitivity with their patients for him in late 1950, and they returned their results to him in early 1951.[55] Their tests showed that women have little vaginal sensitivity, disproving the longstanding Freudian idea that women should be able to transfer their orgasmic sensitivity from their clitorises to their vaginas as an indication that they had reached full adult womanhood. Contrary to Freud, who used no anatomical or physiological evidence to support his claim about orgasmic transfer, the gynecologists found no evidence that women could transfer their nerve endings by any means, either with a man's assistance or through sheer force of will.[56] Kinsey used the gynecologists' tests to criticize a theory based solely on Sigmund Freud and his followers' writings and absent anatomical evidence to support it.[57] Such testing also demonstrated that "the female is not appreciably slower than the male in her capacity to reach orgasm."[58] Kinsey used vaginal sensitivity testing, based as it was on observation, to disprove psychoanalytic theory and to show men's and women's anatomical sexual likeness. Furthermore, he challenged a popular conception of female sexual capacities that women who could not orgasm vaginally were "frigid" and needed psychiatric care.[59]

Kinsey depended on researchers in multiple fields to create the *Female* volume, and his attention to the work of animal researchers is particularly evident in the *Female* volume's chapter on "Physiology of Sexual Response and Orgasm." Ford's and Beach's work on the animal kingdom showed that there was little variation in how animals' bodies changed during the buildup to orgasm. The bodies of humans and primates had analogous physiological manifestations of arousal and orgasm, including higher blood pressure and pulse rates, faster breathing, and loss of sensory capacity, among many other involuntary reactions.[60]

Kinsey asserts that men and women too are similar in how quickly they are aroused and how rapidly they have orgasms. He uses data from anthropological and animal researchers such as Beach, Ford, and Shadle alongside his own research to prove that point. He corrects his previous assertion in the *Male* volume that women take longer to reach orgasm than men.[61] Men could help their female partners have orgasms at their same speed by learning techniques alternate to vaginal intromission alone. In fact, women hardly needed men to help them have orgasms, as his plentiful data on women's masturbation and same-sex behavior to orgasm showed. He suggests to heterosexual men with female partners that they learn female anatomy and physiology, and that they should better understand women's patterns of arousal in order to participate in women's orgasmic pleasure. In fact, "Heterosexual relationships could . . . become more satisfactory if they more often utilized the sort of knowledge which most homosexual females have of female sexual anatomy and female psychology."[62]

Kinsey establishes in the *Female* volume's chapters on anatomy and physiology that men's and women's bodies are remarkably alike in terms of their sexual arousal and responsiveness. In the chapter on psychology, Kinsey puts forth his view that men and women differ in their ability to be conditioned, a view that was in general agreement with Frank Beach's argument on sexual differences in conditionability. Beach and Ford summarize in a comprehensive review of mammalian sexual behavior, "We are strongly impressed with the evidence for sexual learning and conditioning in the male and the relative absence of such processes in the female."[63] Kinsey had many examples of comparative animal behavior that showed that male animals were more responsive to genital, nongenital, contact, and noncontact sexual behavior than female animals were.[64] In the light of those findings, Kinsey argues that men are more susceptible to outside stimuli from adolescence onwards, and that their life experiences make more of an impression on them than women's do on women. Thus men are more likely to have more sexual activity in general and also to have more marginalized sexual behaviors in their histories.

However, Kinsey qualifies Ford and Beach's more sweeping statement about gender difference in conditionability to point out that "it should constantly be borne in mind that there are many individuals, and particularly many females, who widely depart from these averages."[65] Even if women in general were less conditionable regarding their sexual behavior for reasons that scientists had yet to discover, there were too many exceptions for him to make a broad statement about men's and women's psychologies generally. Kinsey concludes in chapter sixteen that men responded in greater numbers than women to all but three of kinds of sexual stimuli: movies, romantic literature, and being bitten. Those three traits as a group had little in common, and Kinsey did not attempt to analyze why more women than men would find them stimulating—he could not attribute the human variation in response to any sexual stimuli to binary gender difference.[66]

As a result of his interest in finding sources for the psychology of sexual behavior, Kinsey and his associates also delved deeply into the fields of neurology and endocrinology. The *Female* volume's sources on how brain chemistry and hormones affected human sexual behavior, and how they affected males and females differently, came from animal behaviorists who had experimented with adding "sex" hormones to laboratory animals. Beginning in the 1920s, endocrinologists experimented with adding or taking away different hormones from animals, usually by castration or ovariectomy and tracking their results. Researchers also cut away portions of animals' brains or glands to see what would happen when they did so. Sometimes both actions were combined: an animal's glands were removed and replaced with artificial hormones, and then its behavior was recorded. Data also existed on the sexual functioning of castrated men and ovariectomized women, usually sex offenders or persons deemed "feeble-minded" by state eugenic law.

Endocrinologists found that reversing the "maleness" and "femaleness" of animals or people by administering hormones, or eliminating the male or female sex drive by surgically removing reproductive organs, was not possible. Androgens and estrogens in manifold combinations are present "in all types of bodies, producing all sorts of different effects," and there was no answer to the question of what, if anything, makes hormones "male" or "female" or what produces "femaleness" or "maleness."[67] There are more hormones in the body than just androgens and estrogens, those hormones govern more than just sexual functions, and hormones are secreted from more places in the human body than just the ovaries and testes. Unfortunately, neither Kinsey nor the specialists on whom he relied for understanding the endocrine system understood much about how and why sex hormones affected the body in the precise ways that they did. Indeed, he comments, "it is unfortunate . . . that these hormones were ever identified as sex hormones, and especially unfortunate that they were identified as male and female sex hormones, for the terminology inevitably prejudices any interpretation of the function of these hormones."[68] Kinsey discusses the findings and implications of his data on the endocrine system's effect on sexual behavior in the *Female* volume's chapter "Hormonal Factors in Sexual Response," but he was unable to isolate the endocrine system's role apart from other bodily systems as they affected sexual behavior.

"Hormonal Factors in Sexual Response," the last chapter of the book, reviews the literature on the relationship of hormones to sexual behavior. Kinsey begins with a discussion of potential lessons that he thought could be drawn from data on hormone response in men and women who had had parts or all of their reproductive organs removed. First, he shows that the removal of the so-called sex glands alone did not stop the sexual impulse; other endocrine functions played key roles in sexual performance. Second, he proves that castrated men and ovariectomized women still had sexual drives and were able to act sexually, thus rendering moot arguments that castration in particular was an effective way to prevent potential future sex crimes. He follows a review of literature on castrated animals and men with the assertion that the data "do not justify the opinion that the public may be protected from socially dangerous types of sex offenders by castration laws."[69]

Kinsey, like the endocrinologists he read, attempted to create a synthetic view of the effect of hormones on male and female sex drives from the data available to him. He was unable to trace specifically which hormones interacted with which other body parts and systems, including the brain. A lack of pituitary hormone decreased sex drive, especially in males, while thyroid deficiencies and overproduction both slowed sexual maturation. Estrogen compromised thyroid function, though testosterone stimulated the thyroid while also increasing metabolism. The liver might secrete a hormone, but some of its secretions somehow inactivated gonadal and possibly other hormones. Paul Robinson argues that Kinsey "eliminated the hormonal explanation for all male–female differences save possibly the aging pattern," even though Kinsey, following Frank Beach's thinking, pointed to the great complexity of the role of hormones in defining sexual behavior and of untangling their relationships to other sex-related factors in the body.[70] For Kinsey, there was no such thing as an explanation for sexual behavior based on a single factor that did not account for the whole complexity of interactive processes within the body and mind.

Throughout the chapter on hormonal factors in sexual response, Kinsey reviews the many studies that physicians and other researchers had completed about the various ways hormones could affect sexual behavior, but none of them was able to make a clear determination of the links—if there were any—of "male" or "female" hormones to any particular type of sexual behavior or identity. Kinsey summarizes his findings on hormones by stating that "while hormonal levels may affect the levels of sexual response . . . there is no demonstrated relationship between any of the hormones and an individual's response to particular sorts of psychological stimuli . . . interest in partners of a particular sex . . . [or] utilization of particular techniques."[71] Kinsey had succeeded in bringing together the most up-to-date material on the relationship of hormones to human sexual behavior, but in the end, even his best effort at synthesis did not answer the question of how that relationship manifested.

Research on the effects of parts of the brain upon sexual behavior would prove slightly less mysterious. Kinsey found that the cerebral cortex was most likely the site where the brain mediated sexual arousal and desire. Beach and other investigators discovered that the cerebral cortex handled most of the learning processes in the animal. Decortication of female rats reduced their sexual functioning to some extent, but decortication of male rats ceased their sexual functioning completely. Beach therefore concluded that the cerebral cortex exerted control over the sexual behavior of males, while females' sexual behavior was controlled by a combination of the cerebral cortex and hormonal mechanisms, with the latter playing the greater role. Cerebral mechanisms could be conditioned, while endocrine glands could not, so it followed that male animals were more "conditionable."

But even if male sexual behavior was marginally more dependent on the cerebral cortex, and hormones influenced female behavior slightly more, the differences between the sexes became less pronounced the larger brains and nervous systems that a species had. Thus any division between men and women regarding their sexual behavior, or between ideas of manhood and womanhood more generally, constituted only a small portion of the total human sexual makeup; their humanness made them more alike than different sexually. Males and females differed quantitatively but not qualitatively in Beach's system. As Anne Fausto-Sterling puts it, "Beach's hypothesis accounted nicely for individual variability within each sex, as well as for the fact that both sexes could, under some conditions, display both masculine and feminine mating patterns and, finally, that both androgen and estrogen could induce either of these patterns in either sex."[72]

Particularly given Lashley's admonishment about the possible errors in attributing response to stimuli directly to an exact area of the brain, Kinsey was reluctant to affirm once and for all that men and women were different as sexed groups. He ends the chapter on psychological responses with the cryptic statement that "the possibilities of reconciling the different sexual interests and capacities of females and males, the possibility of working out sexual adjustments in marriage, and the possibility of adjusting social concepts to allow for these differences . . . will depend upon our willingness to accept the realities which the available data seem to indicate."[73] Perhaps Kinsey wrote this sentence so vaguely because he did not want to leave himself more open to accusations that he was conducting applied research when he claimed to be doing only basic research. He leaves unclear what "social concepts" required adjusting and which "realities" readers should accept in their quest for better marital and social relations. Most likely, the ambiguity of that sentence suggests Kinsey's uneasiness with his decision to make male–female sexual difference a matter of psychological variances in conditionability.

Behavioral sex difference could be observed and quantified, at least according to the comparative measures that Kinsey used, but in all of his and his colleagues' research no one found any concrete evidence of measurable sexed brain difference. At the beginning of the chapter on neurology, he contemplates that "there is nothing yet known in neurologic or physiologic science which explains what we have found" regarding the underlying reasons for their findings. Further on in the chapter, he wrote: "Since we have shown (Chapter 16) that there are considerable differences in the effectiveness of such psychologic stimuli between females and males . . . the most striking disparity which exists between the sexuality of the human female and male, must depend on cerebral differences between the sexes." He continues: "What the nature of such cerebral differences may be, we do not know." He speculates that there is some kind of "mediating mechanism" or a "master switchboard" that controls human sexual arousal and response, and that controls a person's ability to comprehend a given stimulus as sexual.[74] After fifteen years of exhaustive research, the lead author and his research team were only able to say "we do not know" as the answer to a key question about gender difference in sexuality—not just sexual behavior—that had driven the content and scope of the *Female* volume.

Even though Frank Beach's model of gender differences based in neurology made empirical sense to Kinsey, a brain-based model that purported to explain all discrepancies in human sexual behavior had its limits. Kinsey was thus disinclined to examine sociocultural reasons regarding why men's and women's sexual behaviors varied with a similar level of intensive scrutiny, or to consider how sociocultural variations could coexist alongside physical ones. Though the social pressures of gender roles for devout religious women in particular were clear to him, such clarity did not extend to his perceiving the broader patriarchal influences that allowed women limited sexual freedom and agency. For example, Kinsey saw the lack of erotic art and literature created by women, and women's limited response to contemporary erotic art, storytelling, films, and literature, as signs that women were less conditioned psychologically to appreciate and to be aroused by them.[75] He did not consider that the women he studied lived in a patriarchal culture that discouraged them from being active partners in sexual encounters and from free erotic expression (branding such women as prostitutes), and that American culture broadly discouraged them from being sexual subjects instead of objects. Further, as most women had less personal income and less free time than men, especially if they were managing a household, they would have had little time to create their own erotic works, let alone market and sell them.

Kinsey also underestimated unmarried women's concerns about possible pregnancy if they were sexually active with men, in an era when abortion was illegal and often unsafe and an out-of-wedlock pregnancy was a source of embarrassment, shame, and stigma for the pregnant woman. In the 1940s and 1950s, most women had only limited access to forms of birth control in a pre–birth control pill era, none of which were failsafe.[76] Kinsey's dependence on Beach's work, with its brain-based explanation of sexual difference, screened Kinsey's vision from bringing that explanation together with gendered, sociocultural structures of power and agency. As Paul Robinson puts it, "his discussion of male–female differences represented the one serious weakness in an otherwise liberal and humane front. . . . It is [an] enormous tribute to the power of sexual prejudice that [Kinsey] . . . should have articulated a theory of male–female difference that . . . tended to confirm popular opinion."[77]

So even after the most exhaustive review of literature in sexology to date and establishing the largest set of data on human sexual behavior that they could, Kinsey and his team could not definitively say whether sexual similarity or sexual difference was stronger between men and women. After eighteen thousand sex history interviews, thousands of pages of correspondence, and teasing out the possible individual roles of anatomy, physiology, psychology, neurology, and endocrinology in discerning sexual difference, the nature of—and ultimate reasons for—the human categories "male" and "female" as descriptors for human sexual behavior were not good enough. Kinsey repeatedly states throughout the *Female* volume that there are not "sexual qualities which are found only in one or the other sex"; that sexual variation is greater in women than in men; and that "individual variation is the most persistent reality in human behavior."[78]

His data shows the vast, nearly endless amount of variation that exists in human sexual experience, and that data did not fall neatly along gendered lines. He may also have been reluctant to draw gendered lines because of his increasing correspondence and encounters with transsexual and transgender individuals and Harry Benjamin, the leading physician in the United States who worked with them. Though the *Female* volume does not mention transsexuality, and contains only a short discussion of cross-dressing in the chapter on psychology, the idea that men and women could move within and between genders provided Kinsey with even more evidence that two gender categories were inadequate to divide the human race for the discernment of sexual difference.[79]

Throughout the writing, research, editing, and gathering of professional feedback for the *Female* volume, Kinsey came to terms with the fact that he was doing much more than conducting sex history interviews, counting orgasms, working the punched-card machines, and directing his team in making elaborate charts and graphs to illustrate patterns in the masses of Institute for Sex Research data. Understanding sex was about so much more than studying behavior: he wanted to understand and explicate the much broader concept of sexuality—a term he rarely used in print. Perhaps the most accurate understanding of human sexuality as a single entity required the consideration of humans as a single category of analysis.

Conclusion

> I am almost certain that the results of your studies with Cynips, et al., helped to give you further and more profound insight into the possible activities of man, particularly the sexual life of man which . . . plays such an astounding role in man's social and psychological nature.
>
> Edmund E. Jeffers to Alfred C. Kinsey, January 31, 1948

KINSEY'S LIFE AND work, and the intertwined nature of the two, continue to draw academic and public interest. One of the reasons for that ongoing interest is the difficulty in classifying the man himself. Many are intrigued by the highly sexed and voyeuristic Kinsey, who quietly filmed sex acts—his own and those of many others, including his own staff and various sadomasochists—in his attic, while his wife Clara served coffee and persimmon pudding to the tired participants after they put their clothes back on. For others, he fits the image of the scientist in tireless pursuit of truth, working himself to death for the cause. To his harshest critics, he was an abuser of the innocent, proven or not. Sexologists see him as a founder of the field and an inspiration, despite the mistakes that mark his work by present standards. For entomologists, his work has become part of the history of the genus *Cynips*. And for the many people who have found their place on the Kinsey scale, he is an architect who provided them a key to one aspect of their identity. There are other Kinseys, too: the collector of six or seven million gall wasps, the loving husband, the father who gave matter-of-fact sex education to his children, the diligent college and graduate student, the closeted bisexual, the man staying up late to punch cards full of orgasms into big, noisy, punched-card machines. He was the interviewer who got so many people to tell him, a stranger, every fragment of their sexual histories. He wrote the story of those histories in two volumes, *Sexual Behavior in the Human Male* and *Sexual Behavior in the Human Female*, books that irrevocably shaped the gendered and sexualized world in the United States, Western Europe, and beyond. Kinsey, as a person, is difficult to place in one category.

The intention of this book was not, of course, to be a biography of all of these Kinseys, jostling together in literature, film, and people's imaginations. It was rather to draw attention to Kinsey the classifier, who wove his academic life together with organizational techniques, who made the sciences of taxonomy and classification into arts. This book has illustrated how Kinsey's passion for classification structured his scholarship from his earliest graduate career through the *Female* volume. He organized vast collections of all sorts of objects: Mexican oak galls, poisonous and nonpoisonous mushrooms, photographs of women in leather fetish gear, erotic Japanese prints, and the number of orgasms from masturbation per week for unmarried white men between eighteen and twenty-five years of age with a high school education employed as skilled laborers. He classified such a wide range of entities not only for the sake of making small areas of a wide world comprehensible to himself and to his readers; his patterns of classification were also scientific techniques with a broad range of further aims, including to trace the evolution of cynipids in the Northern Hemisphere, to teach high school students to recognize evolution happening all around them, to show residents of the Northeast how to turn "wild" foods into domestic consumables, to look for patterns of how sexual interest manifested in visual culture, and to argue against unjust discrimination regarding the consensual sexual lives of teenagers and adults.

Kinsey's ability to use classification to illustrate and to explain his perception of the natural world on widely different scales, with widely different entities, shows that he could wield a general scientific tool in the service of different ontological purposes. Studying Kinsey's classification of gall wasps, edible wild plants, and human sexual behaviors over his research career reveals how he used techniques of data gathering, organization, and categorizing as part of his efforts to comprehend the natural world. In adhering to his views that naked-eye observation was the best means of ensuring scientific objectivity, he and his assistants produced extensive entomological and sexological works with paper- and machine-based classificatory technologies. He repeatedly affirmed his faith in observation as the method that would come closest to revealing scientific truth.[1] That observation manifested in the careful organizing of mass-collected data, which was studied for patterns in correlation and causation, and then turned into drawings, graphs, charts, and texts through which readers could grapple with the large quantity of data in front of them and the interpretation of the scientists who made it legible.

Kinsey's skill in classification produced the longest-lasting results when he was dismantling hierarchical classifications of entities, not creating or defending them. Though he had many advantages in his own life, in terms of gender, ethnic background, a high level of education, and the social privilege of being a tenured university professor, he thought broadly about how different kinds of advantages and disadvantages structured people's sexual lives and decisions. His 0–6 scale, which initially dismantled an artificial hierarchy between men who were "active" and men who were "receptive" in same-sex sexual activities, re-envisioned the social implications of same- and opposite-sex fantasies, dreams, desires, and behaviors into a scale that placed everyone, regardless of what they had done, were doing, or ever wanted to do, on the same sociocultural plane. The *Female* volume's chapters on anatomy and physiology, which provided pages upon pages of evidence about men's and women's bodily similarity, provided groundwork for second- and third-wave feminist arguments about the sexual equality of men and women. The physiology chapter's dismantling of the myth of the vaginal orgasm, and the concurrent proof that the clitoris was the primary organ of most women's orgasmic satisfaction, further supported later sexual theorists' claims that women had the same rights to, and capacities for, orgasm and sexual pleasure as men. For Kinsey, men and women were still different—his views on psychology saw to that—but those differences need not manifest as differences in social power. After all, "difference . . . is only problematic in a culture when it results in inequality."[2] Shifting the terms by which men and women were classified was a means of shifting their sociocultural power, even though that was not his intention.

Kinsey's attempts at classification reveal the scientist's deep ambition to systematize all of nature, or at least all existing knowledge on a given subject. Such efforts may inevitably fail, or at the very least be unsatisfying, for scholars who hunger to comprehend all of human knowledge and its insights into the natural world. Yet such endeavors by determined individuals are consistently present in the historical record, as few scholars are ever satisfied that they have mastered everything there is to know. Kinsey was one of those scholars. To admit in the *Female* volume's chapter on neurology that he could not answer how the brain ordered sexual response—to say, "we do not know"—may have been as difficult and as distressing as the politically motivated attacks and loss of funding that followed the volume's publication.[3]

Kinsey's classification abilities sometimes fell short of what he aimed to accomplish through their use. Later entomologists revised some of his gall wasp species classifications, and they rejected his and other taxonomists' use of the concept of "varieties" as a subdivision between genus and species. His insistence that orgasm occurs in prepubescent children, his contention that child orgasm be seen on a continuum with adult orgasm, and his use of interview data with child sex offenders alongside data with nonoffenders poisoned his findings for those who deemed sex offender data invalid, if not also morally repugnant. Statisticians rejected his insistence in the *Male* volume that he had proved adolescent men developed patterns in their sex lives corresponding to the social class they would enter into as adults. In the *Female* volume, Kinsey divides the human body into different parts and functions to clarify his analysis of the sexual body as a whole, and to put together the synthetic vision of human sexuality that he craved. After losing his research funding because of the political problems and pressures that his work generated for the Rockefeller Foundation, the need to find adequate funds took too much time away from the research questions—not only what people did, but why and how they did it—that drove his decades-long interest in sexual behavior. The potentialities of a universal concept of sexuality—a concept that explained dreams, desires, behaviors, arousal, and bodily processes on a wide scale—were beyond his grasp. And so it remains for sex researchers and most everyone with an interest in sex beyond their individual experiences.

Nevertheless, Kinsey's efforts at gathering, organizing, and analyzing scientific knowledge added rich new insights to scholarship on gender and sexuality that continue to hold significance through the present. Kinsey's lifelong interest in the organization and classification of knowledge had long-term implications for ideas of gender, sexuality, and for the practice and meaning of classification itself. Kinsey himself wrestled with the idea of human gender even before it became part of the English language. My use of the term "gender" has been ahistorical throughout this book, as it was not a term used to describe characteristics associated with men and women until 1955, when the psychologist John Money adopted it for that purpose.[4] Kinsey's struggles with the ideas of similarity and difference between men and women, what makes an individual "male" or "female," how a scientist might construct an argument about gender, and what evidence a scientist might use for that argument were all demonstrably clear throughout the research and writing of the *Female* volume.

A major difficulty Kinsey had with similarity and difference in gender was in reconciling the similarities of his anatomical and physiological data with the differences expressed to him in sex history interviews with married couples. Married woman after married man told him of the difficulties of correlating their sexual desires and behaviors with those of their spouses, and of how the irreconcilability of such differences often led to separation, divorce, and extramarital petting or intercourse through affairs or with prostitutes. On the micro level, men and women—whether dating or married—had varying degrees of sexual compatibility. On the macro level, men and women appeared to be more alike than different. In both the *Male* and *Female* volumes, Kinsey repeatedly describes the many marriages that were fraught with sexual tension and unhappiness. He argues in the *Female* volume's chapter on psychological differences that male–female sexual relations would improve if husbands and wives realized that their spouses' reactions to their behavior and interests were not individual, but gender typical. In fact, the reverse was true: the variation in sexual behavior was so wide within and between sexes that, for the most "successful" or orgasmic sexual relationships, one needed to pay attention to one's specific partner's needs in each sexual encounter, regardless of gender.[5]

Kinsey's decision to depend on Frank Beach's explanation of sexual differences between men and women as being based in psychology was rooted in the former's faith in scientific observation. Kinsey trusted Beach's observations of animal behavior because they were observations of living creatures copulating without outside interference. He found the evidence of systemic gender inequality across the United States less easy to see. It manifested in ways that he and many other, mostly male, human scientists were unwilling or unable to identify. He would have needed to find a method to observe objectively the often subtle, yet pervasive, discriminations of patriarchal culture that not even his 18,000 sex history interviews, 1,200 literary sources, 70,000 letters, thousands of photographs and artworks, or numerous films of sex acts made in his own attic had revealed to him. He would need to turn an objective, neutral eye toward cultural treatments of gendered wage differences, attitudes toward work, broad cultural stereotypes in religion and media, women's sexual agency, pregnancy, birth control, and a host of other constructions that did not meet his criteria for naked-eye observation. Kinsey was ultimately unable to accept a sociocultural definition of male–female sexual difference because that would mean developing a new understanding of scientific objectivity.

As the *Female* volume developed as a text, it became less about behavior and more about the nature of all human sexuality, within and beyond "male" and "female" as analytical categories. Kinsey used straightforward terms such as "sex," "sexual behavior," "sexual objects," "coitus" ("intercourse" in the *Male* volume), and "masturbation" to catalog and describe the behaviors of his interviewees. He used "sexuality" only in a few circumstances in the *Female* volume when "sexual behavior" seemed not to capture the essence of what he observed. For example, as discussed in chapter six, the two behaviors that merited his use of the term "sexuality" were masturbation and nocturnal sex dreams, as they were two types of behavior that women could engage in without requiring a partner. Masturbation and sexual dreams to orgasm were, in his view, "the activities which provide the best measure of a female's intrinsic sexuality." He also uses the term "sexuality" when, in the passage in that volume's neurology chapter, he admits that he did not know why the differences in male and female sexuality existed.[6] All of the research he had amassed provided no clear answer. By the end of the text, Kinsey was left to grapple with the nature and meaning of the epistemic category of sexuality. He used the term only when he could find no other for a phenomenon that he found difficult to admit was beyond his reach.

Even though grasping the whole of human sexuality as an epistemic concept was an impossible task for Kinsey, the many forms his classification efforts took have lasted into the twenty-first century. Some of those classifications particularly stand out: the 0–6 (heterosexual–homosexual) scale and the popular understanding that 10 percent of the American population is homosexual. The 0–6 scale in the *Male* and *Female* volumes, and the 10 percent figure in the *Male* volume, both reported more than sixty years ago, have become two of the most popular and lasting artifacts of Kinsey's research efforts in American society and culture.[7] In the case of the Kinsey Reports, the hundreds of thousands of data points about sexual behavior produced new relationships between those points via data processing through the punched-card machines. Those data points, transformed into the scale and the 10 percent figure, have served as useful tools for people looking to understand their sexual identity over time and as the groundwork for sexual identity–based activism. Thus Kinsey's data collecting and processing produced new sexual knowledge that continues to affect people's behavior and thinking in the present.

The Kinsey Reports also showed the power of scientific classification to theorize and to explain human sexual behavior to a public—initially American, later worldwide—with a strong interest in comparing their own experiences with those of others, and in understanding the context in which those experiences took place. As Sarah Igo has shown in *The Averaged American*, postwar Americans wanted to know how they fit into, and where they diverged from, the average member of the crowd.[8] The Kinsey Reports made clear that there was no such thing as a wholly "normal" American. Though Robert Yerkes, the chair of the National Research Council, encouraged Kinsey in 1943 to study only normal American teenagers and adults, Kinsey reserved the right to determine the meaning of "normal" on his own terms: "I am very much in sympathy with the idea of getting a picture of the usual portion of the population, before we attempt to analyze the unusual; although such a picture has meant to me, as a taxonomist and a student of variation, securing a picture of the population as a whole."[9] The classification of such wide-ranging personal traits as level of education completed, degree of adherence to religion, penis length, father's profession, and skin color convinced many American readers of the two volumes that they fit many categories of normalcy, but at the same time were also unique individuals whose lives fit into no single category.

The Kinsey Reports also made classification a foundational tool of sexology as a professional field. Although specific journals focused on sex research, such as the *Archive of Sexual Behavior*, were not established until the early 1970s, the Kinsey Reports became a model for how scientists could use classification to support their authority to articulate patterns in, and possible meanings of, human sexual behavior. While other scientists did not follow Kinsey's data-collection patterns or use punched-card machines to process and to organize their findings, his seriousness, comprehensive scholarship, and dedication to the study of human sexual behavior as a science in and of itself laid the foundation for professional sexology. The Kinsey Reports helped future scientists establish "sexual science as a visible and powerful agent of cultural authority over issues of sexuality and gender."[10]

A deep desire to understand the mechanisms that underlay the natural and human worlds motivated Kinsey throughout his research career. Even when he was seriously ill in early July 1956, he thought less of his health and body for his own sake, but more as critical mechanisms necessary to moving his research forward. One month previously, he had attended the dedication ceremonies for Jordan Hall, Indiana University's newest science building at the time, and reconnected with some of his former entomology graduate students, including Albert P. "Pat" Blair. Kinsey wrote to Blair afterward:

> It was very good to have seen you during the dedication program here. It made me proud of all of you, small as the group was, even though it made me regret that things have not worked out so I could continue in that end of the work. There is no question that for me the present work is more imperative.
>
> I spent practically all of June in bed, but there is definite improvement. In spite of a pessimistic doctor I shall prove to them as I have in the past thirty years, that you can do more with a physical handicap than they can sometimes think.[11]

Kinsey, who was not able to overcome the "physical handicap" (an enlarged and weakened heart) that he mentioned in the letter, never lost faith in the ability of basic scientific research to improve the well-being of the environment and humankind, one reader of the gall wasp books or Kinsey Reports at a time.

Kinsey's comprehensive search for all sex-related knowledge to include in his synthetic volumes made him part of a long line of scholars attempting to order and to find meaning in the masses of multimedia information on the human experience. Funding, materials, technologies, time, and health all placed limits on his productivity—as those factors do on any scholar. The American political climate of the 1950s also limited his ability to complete the multivolume project he anticipated, limiting to eighteen thousand the one hundred thousand sex interviews he had eventually planned to collect with his research team. The politics of sex research in the early 1950s led to his funding agency withdrawing due to pressure from Congress (not to mention the FBI, some members of which suspected Kinsey of being a Communist). While he met contemporary standards of scientific objectivity for basic research by his own measures, the funding agency that made the research possible bowed to external pressure to preserve their own reputation and their own access to cultural and governmental power.[12]

In analyzing the history of knowledge, "a crucial historical question . . . is always how knowledge production (as well as transmission) shapes and is shaped by power relations."[13] Kinsey did not have the resources to continue his research without Rockefeller Foundation funding, and thus the potential research that he might have completed if he had not spent much of his time after the publication of the *Female* volume on largely fruitless fundraising efforts was lost. Social epistemologies in every historical era condition attempts to organize knowledge, and in Kinsey's age, the primary shapers were politics, funding, access to organizational technologies, and general interest in scientific sex research.[14] Whether people loved, hated, or were ambivalent about the lead author, the *Male* and *Female* volumes opened conversations on human sexual behavior and human sexuality that continue in academe and in everyday life through the present.

Notes

Introduction

Epigraph: Alfred C. Kinsey, *An Introduction to Biology* (Philadelphia: J. B. Lippincott, 1926), 40.

1. Alfred C. Kinsey, Wardell B. Pomeroy, and Clyde E. Martin, *Sexual Behavior in the Human Male* (Philadelphia: W. B. Saunders, 1948); Alfred C. Kinsey, Wardell B. Pomeroy, Clyde E. Martin, and Paul H. Gebhard, *Sexual Behavior in the Human Female* (Philadelphia: W. B. Saunders, 1953).

2. Kinsey et al., *Sexual Behavior in the Human Female*, 10.

3. Merritt Lyndon Fernald and Alfred Charles Kinsey, *Edible Wild Plants of Eastern North America* (Cornwall-on-Hudson, NY: Idlewild Press, 1943).

4. Jennifer P. Yamashiro, "Sex in the Field: Photography at the Kinsey Institute" (PhD diss., Indiana University, 2002); JoAnn Brooks and Helen C. Hofer, comps., *Sexual Nomenclature: A Thesaurus* (Boston: G. K. Hall & Co., 1976). On Jeanette Howard Foster, the Kinsey Institute's first professional librarian, see Joanne Passet, *Sex Variant Woman: The Life of Jeannette Howard Foster* (Cambridge, MA: Da Capo Press, 2008).

5. For novelists' ideas of Kinsey's life and work, see Irving Wallace, *The Chapman Report* (New York: Simon & Schuster, 1960); T. C. Boyle, *The Inner Circle* (New York: Viking, 2005); for filmmakers' ideas, see *The Chapman Report* (dir. George Cukor, 1962); *Kinsey* (dir. Bill Condon, 2004); and *Kinsey* (dir. Barak Goodman and John Maggio, 2005).

6. Jesse H. Shera, *Libraries and the Organization of Knowledge* (Hamden, CT: Archon Books, 1965).

7. Thomas A. Stapleford, "Market Visions: Expenditure Surveys, Market Research, and Economic Planning in the New Deal," *Journal of American History* 94 (September 2007): 418–44; Thomas A. Stapleford, *The Cost of Living in America: A Political History* (Cambridge, MA: Cambridge University Press, 2009).

8. W. Lloyd Warner and Paul S. Lunt, *The Social Life of a Modern Community* (New Haven, CT: Yale University Press, 1941), 10–11, 8–9.

9. Pitirim Sorokin, *Social Mobility* (New York: Harper & Bros., 1927), iii, 9; "Opinion: Sex or Snake Oil?," *Time*, January 11, 1954, accessed July 26, 2013, http://www.time.com/time/subscriber/article/0,33009,819316-1,00.html.

10. Aron Krich, "Before Kinsey: Continuity in American Sex Research," *Psychoanalytic Review* 53 (Summer 1966): 78.

11. Robert E. Kohler, *All Creatures: Naturalists, Collectors, and Biodiversity, 1880–*

1950 (Princeton, NJ: Princeton University Press, 2006), 229, 266–67; Kohler, *Lords of the Fly:* Drosophila *Genetics and the Experimental Life* (Chicago: University of Chicago Press, 1994), 270.

12. Erika Lorraine Milam, "'The Experimental Animal from the Naturalist's Point of View': Behavior and Evolution at the American Museum of Natural History, 1928–1954," in *Descended from Darwin: Insights into the History of Evolutionary Studies, 1900–1970*, ed. Joe Cain and Michael Ruse, vol. 99, pt. 1 of Transactions of the American Philosophical Society (Philadelphia: American Philosophical Society, 2009), 173.

13. I am grateful to Robert E. Kohler for this insight.

14. Peter Hegarty, *Gentlemen's Disagreement: Alfred Kinsey, Lewis Terman, and the Sexual Politics of Smart Men* (Chicago: University of Chicago Press, 2013), 71.

15. Paul N. Edwards et al., "Historical Perspectives on the Circulation of Information," *American Historical Review* 116 (December 2011): 1398, 1414–15.

16. Noel Malcolm, "Thomas Harrison and His 'Ark of Studies': An Episode in the History of the Organization of Knowledge," *Seventeenth Century* 19 (Autumn 2004): 218. See also Daniel Rosenberg, "Early Modern Information Overload," *Journal of the History of Ideas* 64 (January 2003): 1–9; Peter Burke, *A Social History of Knowledge: From Gutenberg to Diderot* (Cambridge, MA: Polity Press, 2000); Ann Blair, *Too Much to Know: Managing Scholarly Information before the Modern Age* (New Haven, CT: Yale University Press, 2010); and Markus Krajewski, *Paper Machines: About Cards and Catalogs, 1548–1929*, trans. Peter Knapp (Cambridge, MA: MIT Press, 2011).

17. Jonathan Gathorne-Hardy, *Sex the Measure of All Things: A Life of Alfred C. Kinsey* (Bloomington: Indiana University Press, 2004), 218.

18. Cornelia V. Christenson, *Kinsey, A Biography* (Bloomington: Indiana University Press, 1971); Wardell B. Pomeroy, *Dr. Kinsey and the Institute for Sex Research* (New York: Harper & Row, 1972); Pomeroy, *Dr. Kinsey and the Institute for Sex Research*, 2nd ed. (New Haven, CT: Yale University Press, 1982); James H. Jones, *Alfred C. Kinsey: A Public/Private Life* (New York: W. W. Norton, 1997); Gathorne-Hardy, *Sex the Measure of All Things*. For a comparison of those four biographies to other scientific biographies, see James H. Capshew et al., "Kinsey's Biographers: A Historiographical Reconnaissance," *Journal of the History of Sexuality* 12 (July 2003): 465–86.

19. Pomeroy, *Dr. Kinsey and the Institute for Sex Research*, 2nd ed., 336.

20. James H. Jones, "The Origins of the Institute for Sex Research: A History" (PhD diss., Indiana University, 1973), v.

21. Jones, *Alfred C. Kinsey*, 170. For examples of this guilt as a motivating factor for his research, see 22–23, 76, 288, 353, 368, 518–19, and 532. Thomas Laqueur cautions readers to forget "the absence of even the semblance of evidence for Kinsey's inner state" in his review of *Alfred C. Kinsey*. Laqueur, "Sexual Behavior in the

Social Scientist," *Slate*, November 5, 1997, accessed July 19, 2013, http://www.slate.com/articles/arts/books/1997/11/sexual_behavior_in_the_social_scientist.html.

22. Gathorne-Hardy, *Sex the Measure of All Things*, 24, 30, 51, 87, 91, 161.

23. Lynn K. Gorchov, "Sexual Science and Sexual Politics: American Sex Research, 1920–1956" (PhD diss., Johns Hopkins University, 2003); Sarah E. Igo, *The Averaged American: Surveys, Citizens, and the Making of a Mass Public* (Cambridge, MA: Harvard University Press, 2007); Elaine Tyler May, *Homeward Bound: American Families in the Cold War Era* (New York: Basic Books, 1988); Regina Markell Morantz, "Scientist as Sex Crusader: Alfred C. Kinsey and American Culture," *American Quarterly* 29 (Winter 1977): 145–66; Miriam G. Reumann, *American Sexual Character: Sex, Gender, and National Identity in the Kinsey Reports* (Berkeley: University of California Press, 2005).

24. John D'Emilio and Estelle B. Freedman, *Intimate Matters: A History of Sexuality in America* (New York: Harper & Row, 1988); Angus McLaren, *Twentieth-Century Sexuality: A History* (Oxford: Blackwell, 1999); Stephen Garton, *Histories of Sexuality: Antiquity to Sexual Revolution* (New York: Routledge, 2004).

25. Jane Gerhard, *Desiring Revolution: Second Wave Feminism and the Rewriting of American Sexual Thought* (New York: Columbia University Press, 2001); Jennifer Terry, *An American Obsession: Science, Medicine, and Homosexuality in Modern Society* (Chicago: University of Chicago Press, 1999); Paul Robinson, *The Modernization of Sex: Havelock Ellis, Alfred Kinsey, William Masters, and Virginia Johnson* (Ithaca, NY: Cornell University Press, 1989).

26. Julia A. Ericksen, with Sally A. Steffen, *Kiss and Tell: Surveying Sex in the Twentieth Century* (Cambridge, MA: Harvard University Press, 1999); Janice M. Irvine, *Disorders of Desire: Sex and Gender in Modern American Sexology* (Philadelphia: Temple University Press, 1990); Irvine, "From Difference to Sameness: Gender Ideology in Sexual Science," *Journal of Sex Research* 27 (February 1990): 7–23.

27. Joshua P. Levens, "Sex, Neurosis, and Animal Behavior: The Emergence of American Psychobiology and the Research of W. Horsley Gantt and Frank A. Beach" (PhD diss., Johns Hopkins University, 2005); Philip J. Pauly, *Biologists and the Promise of American Life: From Meriwether Lewis to Alfred Kinsey* (Princeton, NJ: Princeton University Press, 2000); Robert E. Kohler, *Landscapes and Labscapes: Exploring the Lab–Field Border in Biology* (Chicago: University of Chicago Press, 2002); Kohler, *All Creatures*.

28. Colin R. Johnson, *Just Queer Folks: Gender and Sexuality in Rural America* (Philadelphia: Temple University Press, 2013), 8.

29. Bruno Latour, *Science in Action: How to Follow Scientists and Engineers through Society* (Cambridge, MA: Harvard University Press, 1987), 227.

30. Robinson, *Modernization of Sex*, 54.

31. Malcolm, "Thomas Harrison and His 'Ark of Studies,'" 217; Latour, *Science in Action*, 254.

32. Kohler, *All Creatures*.

33. Shera, *Libraries and the Organization of Knowledge*, 120.

34. Ibid., 127.

35. Anne Fausto-Sterling, *Sexing the Body: Gender Politics and the Construction of Sexuality* (New York: Basic Books, 2000), 10.

36. Edgar Anderson, "The Problem of Species in the Northern Blue Flags, *Iris versicolor* L. and *Iris virginica* L.," *Annals of the Missouri Botanical Garden* 15 (September 1928): 244; Katharine Bement Davis, *Factors in the Sex Life of Twenty-Two Hundred Women* (New York: Harper & Bros., 1929); Warner and Lunt, *Social Life of a Modern Community* (New Haven, CT: Yale University Press, 1941).

37. Barry Smith and Bert Klagges, "Philosophy and Biomedical Information Systems," in *Applied Ontology: An Introduction*, ed. Katherine Munn and Barry Smith (Frankfurt: Ontos Verlag, 2008), 23; Kinsey et al., *Sexual Behavior in the Human Female*, 385–88.

38. Peter Galison and Lorraine Daston, *Objectivity* (New York: Zone, 2007), 36.

39. Kinsey et al., *Sexual Behavior in the Human Female*, 567.

40. William G. Cochran, Frederick Mosteller, and John W. Tukey, *Statistical Problems of the Kinsey Report on Sexual Behavior in the Human Male* (Washington, DC: American Statistical Association, 1954).

Chapter 1. Learning the Trade, Creating a Collector

Epigraphs: Alfred C. Kinsey to Ralph Voris, March 26, 1931, folder 1, Voris file, Alfred C. Kinsey Correspondence Collection, Kinsey Institute Archives, Bloomington, IN; Karl Sax, "The Bussey Institution: Harvard University Graduate School of Applied Biology, 1908–1936," *Journal of Heredity* 57 (1966): 177.

1. Alfred C. Kinsey, *New Introduction to Biology*, (Chicago: J. B. Lippincott, 1933), vii, 1–2. Untitled notes, 1914–1918, Folder 1: *Amphibolips* Notes, box 1, Alfred C. Kinsey Entomological Papers, American Museum of Natural History (hereafter cited as AMNH) Library, New York, NY; Anne Roe; *The Making of a Scientist* (New York: Dodd, Mead, & Company, 1952), 232.

2. Pauly, *Biologists and the Promise of American Life*, 198–99.

3. William Morton Wheeler, "The Bussey Institution, 1871–1929," in *The Development of Harvard University since the Inauguration of President Eliot 1869–1929*, ed. Samuel Eliot Morison (Cambridge, MA: Harvard University Press, 1930), 508, 516.

4. C. E. Hewitt to Kinsey, April 25, 1931, folder 7, series I.E.2, box I, Alfred C. Kinsey Collection, Kinsey Institute Archives (hereafter cited as KIA); Kinsey to Hewitt, c. 1931, ibid.

5. Fernald and Kinsey, *Edible Wild Plants of Eastern North America*.

6. Sax, "Bussey Institution," 175; J. A. Weir, "Harvard, Agriculture, and the Bussey Institution," *Genetics* 136 (April 1994): 1230; J. A. Weir, "The Bussey Insti-

tution of Harvard University: A Case Study in the History of Agriculture and Genetics," p. 95, unpublished manuscript, 1993, Bussey Institution records, Archives of the Arnold Arboretum of Harvard University (hereafter cited as AAAHU), Jamaica Plain, MA.

7. Sax, "Bussey Institution," 175; Weir, "Harvard, Agriculture, and the Bussey Institution," 1230.

8. Sax, "Bussey Institution," 177; Edgar Anderson to Paul H. Gebhard, "Kinsey As I Knew Him," October 2, 1961, folder 1, series I.F.1, box I, Alfred C. Kinsey Collection, KIA. See also Mary Alice Evans and Howard Ensign Evans, *William Morton Wheeler, Biologist* (Cambridge, MA: Harvard University Press, 1970), 170–71, 195–96; and G. Richard, "The Historical Development of Nineteenth and Twentieth Century Studies on the Behavior of Insects," in *History of Entomology*, ed. Ray F. Smith, Thomas E. Mittler, and Carroll N. Smith (Palo Alto, CA: Annual Reviews, 1973), 489.

9. Karen A. Rader, "'The Mouse People': Murine Genetics Work at the Bussey Institution, 1909–1936," *Journal of the History of Biology* 31 (September 1998): 332, 344–45, 351–52; Sax, "Bussey Institution," 177; Kohler, *Landscapes and Labscapes*, 5, 7; Alfred C. Kinsey, "Life Histories of American Cynipidae," *Bulletin of the American Museum of Natural History* 42 (December 20, 1920): 320–23.

10. Edgar Anderson, interview with Anne Roe, February 1948, p. 5, Edgar Anderson file, Anne Roe Papers, American Philosophical Society (hereafter cited as APS), Philadelphia, PA.

11. Rader, "Mouse People," 332, 344–45, 351–52; Sax, "Bussey Institution," 177; Kohler, *Landscapes and Labscapes*, 5, 7; Kinsey, "Life Histories of American Cynipidae," 320–23.

12. Anderson to Gebhard, "Kinsey As I Knew Him," 1–2.

13. Charlotte Sleigh, *Six Legs Better: A Cultural History of Myrmecology* (Baltimore: Johns Hopkins University Press, 2007), 19.

14. William Morton Wheeler, "The Termitodoxa, or Biology and Society," in *Foibles of Insects and Men* (New York: Alfred A. Knopf, 1928), 210–11, 213, 217. Kinsey, Pomeroy, and Martin, *Sexual Behavior in the Human Male*; Kinsey et al., *Sexual Behavior in the Human Female*; Hegarty, *Gentlemen's Disagreement*, 72–73.

15. Sleigh, *Six Legs Better*, 93.

16. William Morton Wheeler, "The Dry-Rot of Our Academic Biology," in *Foibles of Insects and Men*, 201, 199, 195. For examples of the research Wheeler was criticizing, see Thomas Hunt Morgan, "The Theory of the Gene," *American Naturalist* 51 (September 1917): 513–44; and Calvin B. Bridges and Thomas Hunt Morgan, "The Second-Chromosome Group of Mutant Characters," in *Contributions to the Genetics of Drosophila melanogaster*, 278 (Washington, DC: Carnegie Institute of Washington, 1919): 125–342.

17. Sleigh, *Six Legs Better*, 64.

18. Raymond Pearl, Papers of William Morton Wheeler, Letter to William Morton Wheeler, January 23, 1923, HUGFP 87.10, General Correspondence, 1887–1937, Correspondence 1900–1923 folder, box 39, Harvard University Archives (hereafter cited as HUA), Cambridge, MA; Robert M. Yerkes, Papers of William Morton Wheeler, Letter to William Morton Wheeler, January 23, 1923, ibid.

19. Edward M. East to Richard Goldschmidt, June 17, 1924, E-H Incoming folder, box 2, Richard Benedict Goldschmidt Papers (hereafter cited as RBGP), BANC MSS 72/241 z, Bancroft Library, University of California, Berkeley; East to Goldschmidt, August 23, 1924, ibid. Goldschmidt's side of the correspondence does not survive.

20. Sleigh, *Six Legs Better*, 13.

21. Alfred C. Kinsey, *Methods in Biology* (Chicago: J. B. Lippincott, 1937), 17.

22. However, there is no evidence for the claims that James H. Jones and Jonathan Gathorne-Hardy make that Kinsey's lifelong love of outdoor life was a way to hide or to mitigate his same-sex behavior. Jones, *Alfred C. Kinsey*, 80–84, 152–54, 163, 169–72; Gathorne-Hardy, *Sex the Measure of All Things*, 76, 161–62. Alfred C. Kinsey, *The Origin of Higher Categories in* Cynips (Bloomington: Indiana University Publications, 1936), 15; Kohler, *All Creatures*, 49.

23. Galison and Daston, *Objectivity*, 204.

24. The Phi Beta Kappa address is printed in Christenson, *Kinsey, A Biography*, 3–9.

25. Alfred C. Kinsey, "Course Outline 1928–29," folder 1, series V.A.2.b, box II, Kinsey Collection, KIA; Louise Ritterskamp Rosenzweig, "Reminiscences about Dr. Alfred C. Kinsey," p. 9, c. 1961, folder 22, series I.F.3, box 1, ibid.

26. Sleigh, *Six Legs Better*, 12.

27. Elin K. Jacob, "Classification and Crossdisciplinary Communication: Breaching the Boundaries Imposed by Classificatory Structure," in *Advances in Knowledge Organization, vol. 4, Knowledge Organization and Quality Management*, ed. Hanne Albrechtsen and Susanne Oernager (Frankfurt-am-Main: Indeks Verlag, 1994), 104.

28. Eviatar Zerubavel, *The Fine Line: Making Distinctions in Everyday Life* (Chicago: University of Chicago Press, 1993), 78.

29. Shera, *Libraries and the Organization of Knowledge*, 230.

30. Unnumbered page in notebook, [1908–1921], Papers of William Morton Wheeler, HUGFP 87.65, Teaching Materials: Notes, Manuscripts, and Other Papers, 1886–ca. 1933, box 2, folder 35, "Notebook—Miscellaneous," HUA; Unnumbered page in record book, [1908–1925], box 2, folder 36, "Class record book—Bussey Institution," ibid. Hall, who was a student of the Bussey graduate and Smith College professor Howard M. Parshley, graduated in June 1921 with an ScD in biology; see Howard M. Parshley, Papers of William Morton Wheeler, Letter to William Morton Wheeler, June 5, 1918, HUGFP 87.10, General Correspondence,

1887–1937, P Miscellaneous 1918 folder, box 28, HUA; Wheeler to Parshley, June 7, 1918, ibid; Parshley to Wheeler, March 5, 1919, P Miscellaneous folder 1919, ibid.

31. Weir, "Bussey Institution of Harvard University," AAAHU, 112–16, esp. 112.

32. William Morton Wheeler, "The Classification of Insects," [1909?], 1, 3 (emphasis in original), and 14, Papers of William Morton Wheeler, HUGFP 87.65, Teaching Materials: Notes, Manuscripts, and Other Papers, 1886–ca. 1933, box 1, folder 20, Zoology 7a, "Classification of Insects," HUA.

33. Alfred C. Kinsey, "Lecture Notes of William Morton Wheeler's Course in Entomology [1917–1919]," pp. 5–6, Bussey Institution courses folder, series I.I.2, box I, Kinsey Collection, KIA.

34. Zerubavel, *Fine Line*, 17; Kinsey, "Lecture Notes of William Morton Wheeler's Course in Entomology," KIA, 87, 92; Evans and Evans, *William Morton Wheeler, Biologist*, 193. For comparative work by another Wheeler advisee, see C. L. Metcalf, *The Genitalia of Male Syrphidae* (Columbus: Ohio State University, 1921).

35. Kohler, *All Creatures*, 227.

36. Kinsey, "Gall Wasp Genus *Cynips*," 40–41; Alfred C. Kinsey, "Studies of Gall Wasps (Cynipidae, Hymenoptera)," p. 4, ScD diss., Harvard University, 1919, series I.D.1, box I, Kinsey Collection, KIA.

37. William Morton Wheeler, Papers of William Morton Wheeler, Letter to Vernon Kellogg, February 4, 1931, HUGFP 87.10, General Correspondence, 1887–1937, K Miscellaneous 1931–1932 folder, box 20, HUA.

38. Anderson quoted in Christenson, *Kinsey, A Biography*, 38.

39. Kohler, *All Creatures*, 236.

40. Alfred C. Kinsey to Merritt Lyndon Fernald, March 27, 1920 (emphasis in original), Kinsey Folder, Administrative Correspondence Files, Archives of the Gray Herbarium, Harvard University.

41. Ibid. The phrasing in this letter echoes that in Kinsey's third high school biology textbook: "If, in making a collection, you should discover an undescribed and wholly new insect or plant or other such thing, then you might properly become as excited as Columbus finding a new continent, or Balboa discovering the Pacific Ocean." Alfred C. Kinsey, *New Introduction to Biology*, 2nd ed. (Chicago: J. B. Lippincott, 1938), 92.

42. Kohler, *All Creatures*, 236.

43. Kinsey to Roeth, c. 1920, quoted in Christenson, *Kinsey, A Biography*, 39.

44. Galison and Daston, *Objectivity*, 38.

45. Joe Cain, "Rethinking the Synthesis Period in Evolutionary Studies," *Journal of the History of Biology* 42 (2009): 628–29.

46. Robert Bugbee, quoted in Christenson, *Kinsey, a Biography*, 77. See also Gathorne-Hardy, *Sex the Measure of All Things*, 77.

47. Galison and Daston, *Objectivity*, 40.

48. Anderson to Gebhard, "Kinsey as I Knew Him," 6.

49. June Hiatt Keisler quoted in Christenson, *Kinsey, A Biography*, 73.

50. Gathorne-Hardy, *Sex the Measure of All Things*, 79 (emphasis in original); Cain, "Rethinking the Synthesis Period," 629.

51. Anderson to Gebhard, "Kinsey As I Knew Him," 6.

52. Pomeroy, *Dr. Kinsey and the Institute for Sex Research*, 35; Gathorne-Hardy, *Sex the Measure of All Things*, 469n77.

53. Edgar Anderson, "Hybridization in American *Tradescantias*," *Annals of the Missouri Botanical Garden* 23 (September 1936): 514; see also Kohler, *All Creatures*, 270.

54. Anderson to Gebhard, "Kinsey as I Knew Him," 7–8.

55. Galison and Daston, *Objectivity*, 52.

56. J. A. Kerr and Grove B. Jones, "Soil Survey of Windsor County, Vermont" (Washington, DC, 1918), Alfred C. Kinsey Collection, Rare Map Room, Geography and Map Library, Indiana University, Bloomington; R. A. Winston et al., "Soil Survey of Clearfield County, Pennsylvania" (Washington, DC, 1919), ibid.; Alfred C. Kinsey, "Studies of Some New and Described Cynipidae (Hymenoptera)," *Indiana University Studies* 9 (June 1922): 56, 116; Kinsey, "The Gall Wasp Genus *Neuroterus* (Hymenoptera)," *Indiana University Studies* 10, no. 58 (June 1923): 7.

57. For Kinsey helping an entomologist at the University of Illinois identify a specimen, see W. V. Balduf to Kinsey, Dec. 28, 1925, folder 13: Correspondence: W. V. Balduf, box 1, Kinsey Entomological Papers, AMNH Library; and Kinsey to Balduf, March 26, 1929, ibid. For Kinsey suggesting that another entomologist visit him, see Kinsey to Balduf, January 24, 1927, ibid. For Kinsey organizing a statewide professional meeting, see Alfred C. Kinsey, "INDIANA ENTOMOLOGISTS' MEETING," Nov. 9, 1928, folder 10, ibid. For Kinsey asking to trade specimens with a museum, see Kinsey to Curator of Entomology, Berlin Natural History Museum, March 28, 1928, folder 15: Correspondence: William Beutenmüller and Edna Beutenmüller, ibid.

58. Robert E. Kohler, "*Drosophila* and Evolutionary Genetics: The Moral Economy of Scientific Practice," *History of Science* 29 (1991): 335–75. See also Sleigh, *Six Legs Better*, 15.

59. "American Museum of Natural History, Valuable Gifts," memo, April 25, 1930, folder 1930 (Uncl. Preceding A) A–F, box 1209, 1930–1931, Central Archives, AMNH Library; "American Museum of Natural History, Gifts, 1931, to Date," memo, January 15, 1931, folder 1931, A–K, box 1209, 1930–1931, ibid.; Kohler, *All Creatures*, 123. For Wheeler's relationship with the AMNH, see Evans and Evans, *William Morton Wheeler, Biologist*, 156–57.

60. "American Museum of Natural History, Valuable Gifts," memo, April 25, 1930, folder 1930 (Uncl. Preceding A) A–F, box 1209, 1930–1931, Central Archives, AMNH Library; "American Museum of Natural History, Gifts, 1931, to Date," memo, January 15, 1931, folder 1931, A–K, box 1209, 1930–1931, ibid.; Frank E.

Lutz to G. H. Sherwood, October 16, 1922, folder 53v, Central Archives, AMNH Library. For the correspondence with Banks, see Nathan Banks to Kinsey, June 6, 1923, folder 14: Correspondence: Nathan Banks, box 1, Kinsey Entomological Papers, ibid.; Kohler, *All Creatures*, 229. For Kinsey's donation to the MCZ, see *Annual Report of the Director of the Museum of Comparative Zoology at Harvard College to the President and Fellows of Harvard College for 1922–1923* (Cambridge, MA: Museum of Comparative Zoology, Harvard University, 1923), 7, 14.

61. Paul H. Gebhard to "Division of Insects," February 16, 1960, American Museum of Natural History—Current file, Kinsey Correspondence Collection, KIA and Willis J. Gertsch to Gebhard, February 29, 1960, ibid. See also Mont A. Cazier to Albert E. Parr, memo, Regarding: Kinsey Collection of Gall Wasps as Follows, August 16, 1957, folder 1957, box 1209, 1950–61, Central Archives, AMNH Library; author conversation with Christine LeBeau, Curatorial Assistant, Department of Entomology, AMNH, August 17, 2007.

62. Alfred C. Kinsey, "New Species and Synonymy of American Cynipidae," *Bulletin of the American Museum of Natural History*, 42 (December 20, 1920): 293–317; Kinsey, "Life Histories of American Cynipidae"; Kinsey, "Phylogeny of Cynipid Genera and Biological Characteristics," *Bulletin of the American Museum of Natural History*, 42 (December 20, 1920): 357a–c, 358–402; Kinsey, "Studies of Gall Wasps (Cynipidae, Hymenoptera)," 4.

63. Kinsey, "New Species and Synonymy of American Cynipidae," 301.

64. Kinsey, "Life Histories of American Cynipidae," 319–20, 336–38, 319.

65. Kinsey, "Gall Wasp Genus *Neuroterus* (Hymenoptera)," 8. Stephen Jay Gould, "Of Wasps and WASPS," *Natural History* 91, no. 12 (1982): 10–16.

66. Kohler, *All Creatures*, 242.

67. Alfred C. Kinsey, "Research Report of May 26, 1927," folder 1, series II.A, box I, Kinsey Collection, KIA; Alfred C. Kinsey, "The Gall Wasp Genus *Cynips*: A Study in the Origin of Species," *Indiana University Studies* 84–86 (June, September, December 1929).

68. Kinsey, "Gall Wasp Genus *Cynips*," 15, 458–59; Emmett Reid Dunn, *The Salamanders of the Family Plethodontidae* (Northampton, MA: Smith College, 1926); Clarence Eugene Mickel, "Biological and Taxonomic Investigations on the Mutillid Wasps," [*Smithsonian Institution*] *United States National Museum Bulletin* 143 (1928): 1–351; Anderson, "Problem of Species in the Northern Blue Flags," 244; Jones, *Alfred C. Kinsey*, 222; Gathorne-Hardy, *Sex the Measure of All Things*, 116.

69. William Morton Wheeler, William Morton Wheeler Papers, Letter to Alfred C. Kinsey, March 19, 1930, HUGFP 87.10, General Correspondence, 1887–1937, K Miscellaneous 1930 folder, box 20, HUA; Kinsey, William Morton Wheeler Papers, Letter to Wheeler, March 26, 1930, ibid.

70. Henry Fairfield Osborn, "Aristogenesis, the Creative Principle in the Origin of Species," *Science*, n.s. 79, no. 2038 (January 19, 1934): 44; Kinsey to Henry Fair-

field Osborn, February 23, 1934, Osborn file, Kinsey Correspondence Collection, KIA.

71. Kinsey to Voris, March 26, 1931; Kim Kleinman, "His Own Synthesis: Corn, Edgar Anderson, and Evolutionary Theory in the 1940s," *Journal of the History of Biology* 32, no. 2 (Fall 1999): 301; Kohler, "*Drosophila* and Evolutionary Genetics," 350. See also Donna J. Drucker, "'Building for a life-time of research': Letters of Alfred Kinsey and Ralph Voris," *Indiana Magazine of History* 106 (March 2010): 70–101.

72. Shera, *Libraries and the Organization of Knowledge*, 82.

Chapter 2. The Evolution of a Taxonomist

Epigraph: Alfred C. Kinsey to Ralph Voris, c. October 1939, Ralph Voris file, Kinsey Correspondence Collection, KIA.

1. Peter Del Tredici, "The Other Kinsey Report," *Natural History*, July–August 2006, accessed February 28, 2014, http://www.naturalhistorymag.com/picks-from-the-past/151957/the-other-kinsey-report; Alfred C. Kinsey to Merritt Lyndon Fernald, January 29, 1917, Kinsey folder, Administrative Correspondence Files, Archives of the Gray Herbarium.

2. Del Tredici, "Other Kinsey Report"; Elmer Drew Merrill to Fernald, May 22, 1944, Elmer Drew Merrill Papers, AAAHU.

3. Fernald and Kinsey, *Edible Wild Plants of Eastern North America*, 54.

4. Ibid., 376.

5. Ibid., 2, 5, 7.

6. Ibid., 143, 358.

7. Ibid., 39–40.

8. Kinsey et al., *Sexual Behavior in the Human Female*.

9. Fernald and Kinsey, *Edible Wild Plants of Eastern North America*, vii, viii.

10. Kinsey, Pomeroy, and Martin, *Sexual Behavior in the Human Male*.

11. Alfred C. Kinsey to Merritt Lyndon Fernald, November 13, 1926, Kinsey folder, Administrative Correspondence Files, Archives of the Gray Herbarium.

12. Eric William Engles, "Biology Education in the Public High Schools of the United States from the Progressive Era to the Second World War: A Discursive History" (PhD diss., University of California, Santa Cruz, 1991), 46; Alfred C. Kinsey, *An Introduction to Biology* (Philadelphia: J. B. Lippincott, 1926).

13. Anderson to Gebhard, "Kinsey as I Knew Him," KIA, 1–2.

14. Alfred C. Kinsey, "The Content of the Biology Course," *School Science and Mathematics* 30, no. 4 (April 1930): 375.

15. Johnson, *Just Queer Folks*, 69.

16. Kinsey, "Content of the Biology Course," 382.

17. Alfred C. Kinsey, "Biologic Sciences in Our High Schools," *Proceedings of the Indiana Academy of Science* 35 (1925): 64.

18. Alfred C. Kinsey, *New Introduction to Biology*, xxi–xxiii, xii–xiii.

19. For Kinsey's letter to Wheeler, see Evans and Evans, *William Morton Wheeler, Biologist*, 194.

20. Kinsey, *Introduction to Biology*, x.

21. Ronald Ladouceur, "The Evolution of Textbooks: 1930s Edition," accessed February 28, 2014, http://www.textbookhistory.com/the-evolution-of-textbooks-1930s-edition.

22. Kinsey, *Introduction to Biology*, 204, 209; Kinsey, *New Introduction to Biology*, 231, xv, xii; Ronald Ladouceur, "If Kinsey's Textbook Could Talk . . . ," accessed February 28, 2014, http://www.textbookhistory.com/if-kinsey%e2%80%99s-textbook-could-talk-%e2%80%a6/; Kinsey, *New Introduction to Biology*, 2nd ed., 782–83.

23. Ladouceur, "If Kinsey's Textbook Could Talk."

24. Alfred C. Kinsey, *Field and Laboratory Manual in Biology* (Philadelphia: J. B. Lippincott, 1927), xiv (emphases in original); Kinsey, *Workbook in Biology* (Chicago: J. B. Lippincott, 1934), ix.

25. Sales tallies, folder 1, series IV.C.2, box II, Kinsey Collection, KIA.

26. Constance Areson Clark, "Evolution for John Doe: Pictures, the Public, and the Scopes Trial Debate," *Journal of American History* 87 (March 2001): 1275–1303; Kinsey, *New Introduction to Biology*, 405–18.

27. Kinsey to Edgar Anderson, January 23, 1937, Anderson file, Kinsey Correspondence Collection, KIA; E. W. Bacon to Kinsey, September 10, 1925, J. B. Lippincott file, ibid.; [Jasper?] W. Lippincott to Kinsey, August 19, 1935, ibid.

28. Alfred C. Kinsey, "Suggestions for Revising NEW INTRODUCTION TO BIOLOGY," c. 1941–42, folder 1, series IV.C.4, box II, Kinsey Collection, KIA.

29. Engles, "Biology Education," 124.

30. Kinsey, *Methods in Biology*, 199, 201.

31. Davis, *Factors in the Sex Life of Twenty-Two Hundred Women*; Robert Latou Dickinson and Lura Beam, *A Thousand Marriages: A Medical Study of Sex Adjustment* (Baltimore: Williams & Wilkins, 1931); Robert Latou Dickinson and Lura Beam, *The Single Woman: A Medical Study in Sex Education* (Baltimore: Williams & Wilkins, 1934); Gilbert V. Hamilton, *A Research in Marriage* (New York: Albert and Charles Boni, 1929); William S. Taylor, "A Critique of Sublimation in Males: A Study of Forty Superior Single Men," *Genetic Psychology Monographs* 13 (January 1933): 7–109; Gorchov, "Sexual Science and Sexual Politics," 16.

32. Kinsey, *Methods in Biology*, 200–201.

33. Ibid., 204, 205. See Maurice A. Bigelow, *Sex-Education* (New York: Macmillan, 1916); and John F. W. Meagher and Smith Ely Jelliffe, *A Study of Masturbation and the Psychosexual Life*, 3rd ed. (Baltimore: William Wood, 1936).

34. Kinsey, *Methods in Biology*, 202–4, 25.

35. Bigelow, *Sex-Education*, 43–45, 59.

36. For examples of Rice's writing for young people, see Thurman B. Rice,

The Age of Romance (Chicago: American Medical Association, 1933), *How Life Goes On and On: A Story for Girls of High School Age* (Chicago: American Medical Association, 1933), and *The Human Body: Some Rules for Right Living* (New York: Funk & Wagnalls, 1937).

37. [Indiana] Committee on Secondary School Science Curriculum, "Preliminary Statement: Biology in the High School Curriculum," p. 2, February 1943, folder 5, series V.C., box II, Kinsey Collection, KIA. See also "Minutes of American Association for the Advancement of Science in General Education," April 30–May 1, 1938, Columbus, Ohio, folder 4, ibid.; Kinsey, spring 1932 final exam, May 31, 1932, folder 3, series V.A.6, box II, ibid.; "Dr. Kinsey to Conduct Round Table Discussion on High School Biology," *Indiana Daily Student*, July 30, 1938, 4; "Faculty Members to Give Talks at Extension Center," *Indiana Daily Student*, February 4, 1939, 2; L. W. Taylor, R. W. Taylor, and Alfred C. Kinsey, "An Invitation to Teachers of Science," *Science* n.s., 87, no. 2267 (June 10, 1938): 528.

38. Kinsey, "Gall Wasp Genus *Cynips*"; Kinsey, *Origin of Higher Categories in* Cynips.

39. Cain, "Rethinking the Synthesis Period in Evolutionary Studies," 625n1, 626. Others stating the need for a broader synthesis include Juan Ilerbaig, "'The View-Point of a Naturalist': American Field Zoologists and the Evolutionary Synthesis, 1900–1945," in Cain and Ruse, *Descended from Darwin*, 33; Kim Kleinman, "Biosystematics and the Origin of Species: Edgar Anderson, W. H. Camp, and the Evolutionary Synthesis," in ibid., 81; Gregory K. Davis, Michael R. Dietrich, and David K. Jacobs, "Homeotic Mutants and the Assimilation of Developmental Genetics into the Evolutionary Synthesis, 1915–1952," in ibid., 135.

40. Richard Goldschmidt, *The Material Basis of Evolution* (New Haven, CT: Yale University Press, 1940); Nikolai Krementsov, "A Particular Synthesis: Aleksander Promptov and Speciation in Birds," *Journal of the History of Biology* 40 (2007): 637–82; Kleinman, "His Own Synthesis," 297; Anderson, "Problem of Species in the Northern Blue Flags"; Edgar Anderson, "Supra-specific Variation in Nature and in Classification from the View-Point of Botany," *American Naturalist* 71 (1937): 223–35.

41. See Mont A. Cazier to Albert E. Parr, memo, Regarding: Kinsey Collection of Gall Wasps as Follows, August 16, 1957, folder 1957, box 1209, 1950–1961, Central Archives, AMNH Library; Kohler, *Landscapes and Labscapes*, 303.

42. Kinsey to Voris, September 3, 1934, Voris file, Kinsey Correspondence Collection, KIA; see also Drucker, "Building for a life-time of research," 70–101.

43. Kinsey, *Origin of Higher Categories in* Cynips, 5. For a review, see G. S. Walley, review of *The Origin of Higher Categories in* Cynips, by Alfred C. Kinsey, *Canadian Entomologist* 69 (January 1937): 20–21.

44. Kinsey, *Origins of Higher Categories in* Cynips, 63.

45. Ibid., 60, 8–14; Alfred C. Kinsey, "Supra-specific Variation in Nature and in Classification from the View-Point of Zoology," *American Naturalist* 71 (May–June

1937): 218–19; Kohler, *All Creatures*, 243, 246; Cain, "Rethinking the Synthesis Period in Evolutionary Studies," 36.

46. Anderson to Kinsey, November 17, 1936, Anderson file, Kinsey Correspondence Collection, KIA.

47. George Gaylord Simpson to William King Gregory, November 18, 1936, Series I, George Gaylord Simpson Papers, APS.

48. George Gaylord Simpson, "Supra-specific Variation in Nature and in Classification from the View-Point of Paleontology," *American Naturalist* 71 (May–June 1937): 236. Simpson later concluded that this conference paper was one of two that prepared him to write his own contribution to the evolutionary synthesis, *Tempo and Mode in Evolution*. In a roundabout way, Kinsey encouraged Simpson to think more broadly about evolutionary theory so that his own theories would be more broadly applicable to species across time. As Kinsey used poor studies of sex education and masturbation to clarify his thinking, so too did Simpson use Kinsey's incorrect scholarship to sharpen his own. Léo F. Laporte, "George G. Simpson, Paleontology, and the Expansion of Biology," in *The Expansion of American Biology*, ed. Keith R. Benson, Jane Maienschein, and Ronald Rainger (New Brunswick, NJ: Rutgers University Press, 1991), 85–86; George Gaylord Simpson, *Tempo and Mode in Evolution* (New York: Columbia University Press, 1944).

49. Simpson, "Supra-specific Variation in Nature and in Classification from the View-Point of Paleontology," 250.

50. Joseph Allen Cain, "Common Problems and Cooperative Solutions: Organizational Activity in Evolutionary Biology, 1936–1947," *Isis* 84 (March 1993): 3; Simpson, "Supra-specific Variation in Nature and in Classification from the View-Point of Paleontology," 253, 255, 265, 267; Kinsey, *Origin of Higher Categories in Cynips*, 48, 66.

51. Richard Goldschmidt, "*Cynips* and *Lymantria*," *American Naturalist* 71 (September–October 1937): 508; Ernst Mayr, "The Role of Systematics in the Evolutionary Synthesis," in *The Evolutionary Synthesis: Perspectives on the Unification of Biology*, ed. Ernst Mayr and William B. Provine (Cambridge, MA: Harvard University Press, 1980), 128; Gould, "Of Wasps and WASPS," 10–16.

52. George Gaylord Simpson to Kinsey, January 31, 1938, Series I, Simpson Papers, APS; Kinsey to Simpson, February 17, 1938, ibid.; Simpson to Kinsey, February 23, 1938, ibid. Kinsey to Anderson, January 12, 1937, Anderson file, Kinsey Correspondence Collection, KIA; Kinsey to Voris, January 31, 1937, Voris file, Kinsey Correspondence Collection, KIA; Kinsey to Voris, November 14, 1937, ibid.; Voris to Kinsey, November 17, 1937, ibid.; Kohler, *All Creatures*, 229; Cain, "Rethinking the Synthesis Period in Evolutionary Studies," 639. Kinsey also received a star by his entry in the 1937 edition of *American Men of Science*, a periodical listing the most prominent scientists of the year. The Kinsey Institute, "Alfred Kinsey Collection," accessed February 28, 2014, http://www.indiana.edu/~kinsey/library/kinsey.html.

53. Richard M. Burian, "How the Choice of Experimental Organism Matters: Epistemological Reflections on an Aspect of Biological Practice," *Journal of the History of Biology* 26 (Summer 1993): 357.

54. Sleigh, *Six Legs Better*, 70–71.

55. Vassiliki Betty Smocovitis, *Unifying Biology: The Evolutionary Synthesis and Evolutionary Biology* (Princeton, NJ: Princeton University Press, 1996), 56.

56. Burian, "How the Choice of Experimental Organism Matters," 354, 361–62.

57. Cain, "Rethinking the Synthesis Period in Evolutionary Studies," 639.

58. Alfred E. Emerson, "Objects of the Society for the Study of Speciation" [1940], p. 1, History: Objects of the SSS & Membership List [1940] folder, Series VIII, Society for the Study of Evolution Papers, APS. For a more detailed discussion of the formative years of the SSS and the Society for the Study of Evolution (SSE), see Vassiliki Betty Smocovitis, "Organizing Evolution: Founding the Society for the Study of Evolution (1939–1950)," *Journal of the History of Biology* 27 (Summer 1994): 241–309.

59. Emerson, "Objects of the Society for the Study of Speciation," APS, 14, 12, 15–16.

60. Kinsey, Pomeroy, and Martin, *Sexual Behavior in the Human Male*.

61. "The Society for the Study of Evolution Organizational Meeting, March 30, 1946, St. Louis, Missouri," 1946, Minutes. Organizational Meeting 1946 March 30 folder, series 1, box 1, Society for the Study of Evolution Papers, APS; Ernst Mayr, "History of the Society for the Study of Evolution," 1952, pp. 1–3, "History. History of SSE by E. Mayr" folder, Series VIII, ibid.; Joe Cain, ed., "Exploring the Borderlands: Documents of the Committee on Common Problems of Genetics, Paleontology, and Systematics, 1943–1944," *Transactions of the American Philosophical Society* 94 pt. 2 (2004): 113; Smocovitis, "Organizing Evolution," 263–64.

62. Anderson to Kinsey, March 18, 1949, Anderson file, Kinsey Correspondence Collection, KIA; Kinsey to Anderson, April 2, 1949, ibid.; Ernst Mayr to Kinsey, May 12, 1944, American Museum of Natural History file, ibid.; Kinsey to Mayr, May 16, 1944, ibid.; Kinsey to J. B. S. Haldane, December 23, 1948, J. B. S. Haldane file, ibid.; Haldane to Kinsey, January 22, 1949, ibid.; Haldane to Kinsey, May 19, 1954, ibid.; Kinsey to Haldane, July 14, 1954, ibid.; Haldane to Kinsey, January 13, 1955, ibid.; Kinsey to Haldane, February 1, 1955, ibid.; Haldane to Kinsey, May 31, 1955, ibid.; Kinsey to Haldane and Helen Spurway Haldane, December 19, 1955, ibid.; Haldane to Kinsey, January 1, 1956, ibid.; Haldane to Wardell B. Pomeroy, July 9, 1957, ibid.; Pomeroy to Haldane, July 25, 1957, ibid.; Pomeroy to Haldane, January 28, 1958, ibid.; Haldane to Pomeroy, February 4, 1958, ibid.; Pomeroy to Haldane, February 11, 1958, ibid.; Pomeroy to Haldane, January 12, 1959, ibid.

63. Cain, "Rethinking the Synthesis Period in Evolutionary Studies," 627n3.

Chapter 3. Teaching Life and Human Sciences

Epigraph: Kinsey to Voris, July 6, 1939, Voris file, Kinsey Correspondence Collection, Kinsey Institute Archives.

1. Kinsey, "Lecture Notes of William Morton Wheeler's Course in Entomology," KIA, 5–6.
2. Kinsey, Pomeroy, and Martin, *Sexual Behavior in the Human Male*; Kinsey et al., *Sexual Behavior in the Human Female*.
3. Galison and Daston, *Objectivity*, 326.
4. Alfred C. Kinsey, "Course Outline," folder 11, box 1, Alfred C. Kinsey Entomological Papers, AMNH Library.
5. Alfred C. Kinsey, "Annual Report April 12, 1941," folder 8, series II.A, box II, Kinsey Collection, KIA; Alfred C. Kinsey, evolution class notes, pp. 1–2, folder 1, series V.A.2.a, ibid. Charles Darwin, *On the Origin of Species by Means of Natural Selection* (London: John Murray, 1859); Darwin, *The Descent of Man, and Selection in Relation to Sex* (London: John Murray, 1871); Thomas Hunt Morgan, *The Scientific Basis of Evolution* (New York: W. W. Norton, 1932); Henry Fairfield Osborn, *The Origin and Evolution of Life: On the Theory of Action, Reaction, and Interaction of Energy* (New York: C. Scribner, 1925).
6. Alfred C. Kinsey, "Lectures–Biology" notebook, fall 1922, pp. 1–2, 7, folder 1, series V.A.2.a, box II, Kinsey Collection, KIA; Alfred C. Kinsey, spring 1932 final exam, May 31, 1932, folder 3, series V.A.6, ibid.
7. For examples, see Kinsey, *Workbook in Biology*, 175, 185, 188, 193, 209; Kinsey, *Methods in Biology*, vii–viii, 4–5, 24, 27, 71.
8. Smocovitis, *Unifying Biology*, 115.
9. Galison and Daston, *Objectivity*, 327.
10. Alfred C. Kinsey, evolution class notes, p. 1, c. 1935, folder 1, series V.A.7, box II, Kinsey Collection, KIA; Dunn, *Salamanders of the Family Plethodontidae*; Sewall Wright, "The Role of Mutation, Inbreeding, Crossbreeding, and Selection in Evolution," *Proceedings of the Sixth International Congress of Genetics* 1 (1932): 356–66; Theodosius Dobzhansky, "A Critique of the Species Concept in Biology," *Philosophy of Science* 2 (July 1935): 344–55; Dobzhanksy, *Genetics and the Origin of Species* (New York: Columbia University Press, 1937).
11. Kinsey, evolution class notes, pp. 45, 18, 74, 59, and 55.
12. Kinsey's 1926 letter to William Morton Wheeler reprinted in Evans and Evans, *William Morton Wheeler, Biologist*, 194.
13. Alfred C. Kinsey, evolution final exam, December 17, 1943, folder 1, series V.A.7, box II, Kinsey Collection, KIA; Alfred C. Kinsey, evolution exam, October 11, [1946?], ibid.

14. "I.U. to Offer Course in 'Marriage': In Form of Series of Lectures and Will Be Open to Selected Group," *Bloomington Daily Telephone*, June 23, 1938, 1.

15. Garton, *Histories of Sexuality*, 200; James H. Capshew, *Herman B Wells: The Promise of the American University* (Bloomington: Indiana University Press, 2012), 118–20; Wheeler, "Termitodoxa, or Biology and Society," 205–17; Alfred C. Kinsey to Ralph Voris, c. October 1939, Voris file, Kinsey Correspondence Collection, KIA.

16. "University Physician Favors Student Wasserman Tests," *Indiana Daily Student*, February 15, 1938, 1; "For a Progressive Indiana," *Indiana Daily Student*, April 17, 1938, 4; Allan M. Brandt, *No Magic Bullet: A Social History of Venereal Disease in the United States since 1880* (New York: Oxford University Press, 1985), 147–49.

17. "Enrollment Figures Show All-Time High of 6,106 Students at Indiana University," *Indiana Daily Student*, October 8, 1938, 1; Jennie Posillico, "Collegiana," *Indiana Daily Student*, October 13, 1937, 4; Jennie Posillico, "Scannin' Collegiana," *Indiana Daily Student*, February 12, 1938, 4; "Dr. Thurman B. Rice to Speak at Meet: I. U. Medical Professor to Lead Discussions on Marriage at Summer Conference," *Indiana Daily Student*, May 28, 1938, 3; "Mrs. J. C. Todd to Discuss 'Choice of Life Partner,'" *Indiana Daily Student*, January 14, 1939, 3; "Kohlmeier to Discuss 'Marriage Philosophy,'" *Indiana Daily Student*, February 4, 1939, 4; "Church Forum to Hear Talk by Rev. Moore," *Indiana Daily Student*, February 18, 1939, 3; Davis, *Factors in the Sex Life of Twenty-Two Hundred Women*, 66–67; Julian B. Carter, "Birds, Bees, and Venereal Disease: Toward an Intellectual History of Sex Education," *Journal of the History of Sexuality* 10 (April 2001): 244–45; letter to the editor, *Indiana Daily Student*, March 8, 1938, 4; letter to the editor, *Indiana Daily Student*, October 31, 1939, 4; letter to the editor, *Indiana Daily Student*, November 8, 1939, 4.

18. Board of Trustees Minutes, June 9, 1938, vol. II, 259–60, Indiana University Archives (hereafter cited as IUA), Bloomington, IN; Christine Carlson et al. to Herman B Wells, May 14, 1938, Kinsey-Marriage Course file, Wells Papers, IUA; "I.U. to Offer Course in 'Marriage,'" 8; Herman B Wells, *Being Lucky: Reminiscences and Reflections* (Bloomington: Indiana University Press, 1980), 100; Clara McMillen Kinsey, interview by James H. Jones, typescript, December 10, 1971, Center for the Study of History and Memory (hereafter cited as CSHM), Indiana University, Bloomington, IN; Kate Hevner Mueller, interview by James H. Jones, p. 18, typescript, April 1, 1971, CSHM; Kate Hevner Mueller to Alfred C. Kinsey, October 25, 1938, Kate Heyner Mueller file, Kinsey Correspondence Collection, KIA; Beth Bailey, "Scientific Truth . . . and Love: The Marriage Education Movement in the United States," *Journal of Social History* 20 (Summer 1987): 718–20; Jones, *Alfred C. Kinsey*, 322, 326, 828n36, 829n53; Gathorne-Hardy, *Sex the Measure of All Things*; 124–25, 151, 473n2. O. F. Hall to Alfred C. Kinsey, April 15, 1938, Hall file, Kinsey Correspondence Collection, KIA; Norman E. Himes to Alfred C. Kinsey, April 22, 1938, Himes file, ibid.; Kinsey to Himes, April 29, 1938, ibid.; Kinsey to Himes, May

Notes to Pages 69–71 **189**

7, 1938, ibid.; William G. Mather Jr. to Alfred C. Kinsey, May 12, 1938, Mather file, ibid.; Herman B Wells to Fowler Harper et al., July 9, 1938, Wells file, ibid.; Kinsey to Wells, July 19, 1938, ibid.; Wells to Kinsey, July 21, 1938, ibid.; Edith Schuman, interview by James H. Jones, typescript, September 15, 1971, CSHM; Kate Mueller, interview; Alfred C. Kinsey, "Indiana University Marriage Course—Fall 1938," c. November 1938, folder 1, series V.A.1.J, box II, Kinsey Collection, KIA; Harvey J. Locke, "Outline of Family Disorganization," July 28, 1938, Marriage Course 1942–43 Folder (9081–25), John H. Mueller Papers, IUA.

19. Kinsey recommended the following for the marriage course faculty: Davis, *Factors in the Sex Life of Twenty-Two Hundred Women*; Katharine Bement Davis, "Periodicity of Sex Desire," *American Journal of Obstetrics and Gynecology* 12 (December 1926): 824; Gerrit S. Miller Jr., "The Primate Basis of Human Sexual Behavior," *Quarterly Review of Biology* 6 (December 1931): 379–410; Olga Knopf, *The Art of Being a Woman*, trans. Alan Porter (Boston: Little, Brown, 1932); William S. Taylor, *A Critique of Sublimation in Males* (Worcester, MA: Clark University, 1933); Robert Briffault, *The Mothers: A Study of the Origins of Sentiments and Institutions* (New York: Macmillan, 1931); Paul Popenoe, *Problems of Human Reproduction* (Baltimore: Williams & Wilkins, 1926); Havelock Ellis, *Psychology of Sex: A Manual for Students* (New York: Ray Long & Richard R. Smith, 1933); Eric M. Matsner, *The Technique of Contraception* (Baltimore: William Wood, 1936). Davis's "The Periodicity of Sex Desire" was reprinted in her *Factors in the Sex Life of Twenty-Two Hundred Women*, 187–217. Taylor's book was first published as "A Critique of Sublimation in Males: A Study of Forty Superior Single Men," *Genetic Psychology Monographs* 13 (January 1933): 7–109. Marriage course faculty reading list, c. May 1938, folder 1, series V.A.1.k, box II, Kinsey Collection, KIA.

20. Davis, *Factors in the Sex Life of Twenty-Two Hundred Women*, 76–77. For a comparison of Davis with other highly educated female Progressive Era reformers, see Ellen F. Fitzpatrick, *Endless Crusade: Women Social Scientists and Progressive Reform* (New York: Oxford University Press, 1990).

21. Peter Hegarty, "Getting Miles Away from Terman: Did the CRPS Fund Catharine Cox Miles's Unsilenced Psychology of Sex?" *History of Psychology* 15, no. 3 (2012): 207.

22. Taylor, "Critique of Sublimation in Males," 8, 13.

23. Ibid., 33.

24. Ibid., 7.

25. Alfred C. Kinsey, "Reproductive Anatomy and Physiology," p. 1, July 12, 1938, folder 1, series V.A.1.b, box II, Kinsey Collection, KIA. Bronislaw Malinowski, *The Sexual Life of Savages in North-Western Melanesia: An Ethnographic Account of Courtship, Marriage, and Family Life among the Natives of the Trobriand Islands, British New Guinea* (New York: Halcyon House, 1929); Hannah M. Stone and Abraham

Stone, *A Marriage Manual: A Practical Guide-Book to Sex and Marriage* (New York: Simon & Schuster, 1935).

26. "Summary of Student Answers," November 1938, folder 5, series V.A.1.i, box II, Kinsey Collection, KIA; "Indiana University Marriage Course Case Histories," c. November 1938, folder 1, series V.A.1.m, box II, Kinsey Collection, KIA.

27. Stone and Stone, *Marriage Manual*, v.

28. Malinowski, *Sexual Life of Savages in North-Western Melanesia*, 67, 69.

29. Miller, "Primate Basis of Human Sexual Behavior," 385; Howard M. Parshley, *The Science of Human Reproduction* (New York: Eugenics Publishing Co., 1933), 282. Gilbert V. Hamilton, "A Study of Sexual Tendencies of Monkeys and Baboons," *Journal of Animal Behavior* 4, no. 5 (1914): 295–318.

30. Miller, "Primate Basis of Human Sexual Behavior," 386, 390, 398; Parshley, *Science of Human Reproduction*, 294.

31. Lester Frank Ward, *Pure Sociology* (New York: Macmillan, 1903), 296–373. See also Judith A. Allen, "'The Overthrow' of Gynaecocentric Culture: Charlotte Perkins Gilman and Lester Frank Ward," in *Studies in American Literary Realism and Naturalism (Salrn)*, ed. Cynthia J. Davis and Denise D. Knight (Tuscaloosa: University of Alabama Press, 2004), 59–86.

32. The sex history interview questions about sex acts imply that they were done with consent, but the interviewers do not specifically ask about consent. The interview team did not ask interviewees about violence used by or against them related to sex acts, unless they were in prison for sex offenses. Joan Scherer Brewer, ed., *The Kinsey Interview Kit* (Bloomington: Kinsey Institute for Research in Sex, Gender, and Reproduction, 1985), 86–87, 100, 130.

33. Robert Latou Dickinson, *Human Sex Anatomy: A Topographical Hand Atlas* (Baltimore: Williams & Wilkins, 1933); Jones, *Alfred C. Kinsey*, 831n18. Several of these drawings were reproduced in Parshley's book on human reproduction, where Kinsey may have found them first. See Parshley, *Science of Human Reproduction*.

34. Kinsey to Dickinson, October 30, 1933, folder 8a, series IIc, box 1, Robert Latou Dickinson Papers, KIA; Kinsey to Dickinson, June 23, 1941, ibid.; Kinsey to Dickinson, February 3, 1943, ibid.; Kinsey to Dickinson, September 5, 1950, folder 8c, ibid.

35. Kinsey to Dickinson, October 30, 1933; Alfred C. Kinsey, "Criteria for a Hormonal Explanation of the Homosexual," *Journal of Clinical Endocrinology* 1 (May 1941): 424–28; Kinsey to Dickinson, June 23, 1941; Kinsey to Dickinson, February 3, 1943; Kinsey to Dickinson, January 4, 1944, folder 8a, series IIc, box 1, Dickinson Papers, KIA; Kinsey, Pomeroy, and Martin, *Sexual Behavior in the Human Male*, viii.

36. Jones, *Alfred C. Kinsey*, 831n18.

37. Dickinson, *Human Sex Anatomy*, 119.

38. Ibid., vii; Gorchov, "Sexual Science and Sexual Politics," 14–16.

39. Kinsey, *Methods in Biology*, 203–4.

40. Alfred C. Kinsey, "Biologic Bases of Society," pp. 10–11, June 28, 1938, folder 1, series V.A.1.g, box II, Kinsey Collection, KIA; Alfred C. Kinsey, "First Lecture of Marriage Course," p. 11, June 15, 1939, folder 1, series V.A.1.d, box II, Kinsey Collection, KIA; Alfred C. Kinsey, "Bases of Society," p. 11, February 5, 1940, folder 1, series V.A.1.e, box II, Kinsey Collection, KIA.

41. Briffault, *Mothers*; Knopf, *Art of Being a Woman*.

42. "Course in Marriage," *Indiana Daily Student*, June 22, 1938, 3; "Summary of Student Answers," August 1938, folder 1, series V.A.1.i, box II, Kinsey Collection, KIA.

43. Kinsey, "Biologic Bases of Society," KIA, 1, 7, 11; Parshley, *Science of Human Reproduction*, 300–301, 304; Davis, *Factors in the Sex Life of Twenty-Two Hundred Women*, 65; Raymond Squier, "The Medical Basis of Intelligent Sexual Practice," in *Plan for Marriage: An Intelligent Approach to Marriage and Parenthood*, ed. Joseph Kirk Folsom (New York: Harper & Bros., 1938), 125, 127–28; Phyllis Blanchard and Carlyn Manasses, *New Girls for Old* (New York: Macaulay, 1937), 191. On debates over sex before marriage and the phenomenon of "trial marriage" in the 1920s, see Rebecca L. Davis, "'Not Marriage at All, But Simple Harlotry': The Companionate Marriage Controversy," *Journal of American History* 94 (March 2008): 1137–63.

44. Squier, "Medical Basis of Intelligent Sexual Practice," 119; Alfred C. Kinsey, "Reproductive Anatomy and Physiology," July 12, 1938, p. 13, folder 1, series V.A.1.b, box II, Kinsey Collection, KIA; Stone and Stone, *Marriage Manual*, 158–59; Davis, *Factors in the Sex Life of Twenty-Two Hundred Women*, 66–67, 69; Carter, "Birds, Bees, and Venereal Disease," 238–39, 247; Jessamyn Neuhaus, "The Importance of Being Orgasmic: Sexuality, Gender, and Marital Sex Manuals in the United States, 1920–1963," *Journal of the History of Sexuality* 9 (April 2000): 456–57.

45. Squier, "Medical Basis of Intelligent Sexual Practice," 129–30; Robert L. Kroc, "Endocrine Basis of Sex and Reproduction," pp. 10, 12, July 14, 1938, folder 2, series V.A.1.b, box II, Kinsey Collection, KIA; Fausto-Sterling, *Sexing the Body*, 182–83.

46. Alfred C. Kinsey, "Individual Variation," July 19, 1938, pp. 1, 5, folder 3, series V.A.1.b, box II, Kinsey Collection, KIA; Squier, "Medical Basis of Intelligent Sexual Practice," 120; Stone and Stone, *Marriage Manual*, 172–73; Neuhaus, "Importance of Being Orgasmic," 450, 457–58; Blanchard and Manasses, *New Girls for Old*, 196, 198; Paul Popenoe, *Preparing for Marriage* (Los Angeles: American Institute of Family Relations, 1938), 10–12. One of the first participants in the marriage course interviewed for a Kinsey Institute documentary remembered Kinsey's penis–clitoris comparisons specifically. See *Sex and the Scientist* (dir. Jim Bales, 1989).

47. Harvey J. Locke, "Family Disorganization," pp. 1, 9, July 28, 1938, folder 6, series V.A.1.b, box II, Kinsey Collection, KIA.

48. "Summary of Student Answers," August 1938. Alfred C. Kinsey to Herman B Wells, September 12, 1938, Wells file, Kinsey Correspondence Collection, KIA.

49. Pomeroy, *Alfred Kinsey and the Institute for Sex Research*, 63.

50. Ernest R. Groves and William F. Ogburn, *American Marriage and Family Relationships* (New York: Henry Holt, 1928); *Plan for Marriage*, ed. Folsom. For examples of marriage course instructors using the individual sex conference, see J. Stewart Burgess, "The College and the Preparation for Marriage and Family Relationships," *Living* 1 (May–August 1939): 39–42; Mary A. Johnson, "A Course in Human Relations at Brooklyn College," *Living* 1 (November 1939): 73–74; Moses Jung, "The Course in Modern Marriage at the State University of Iowa," *Living* 1 (May–August 1939): 43, 50; Bailey, "Scientific Truth," 721–22. For Groves's and Bowman's use of the individual sex conference, see Henry Bowman, "The Marriage Course at Stephens College," *Marriage and Family Living* 3 (February 1941): 8–9, 11; Henry Bowman, Flora Thurston, and Margaret Wylie, "The Teacher as Marriage and Family Counselor," *Marriage and Family Living* 6 (November 1944): 76–78; Ernest R. Groves, *Marriage* (New York: Henry Holt, 1933), x.

51. Alfred C. Kinsey to marriage course staff, January 11, 1939, folder 2, series V.A.1.j, box II, Kinsey Collection, KIA; Alfred C. Kinsey to Ralph Voris, November 28, 1938, Voris file, Kinsey Correspondence Collection, KIA; Kinsey to Voris, January 17, 1939, ibid.; Pomeroy, *Dr. Kinsey and the Institute for Sex Research*, 72.

52. "Summary of Student Answers," November 1938; "Indiana University Marriage Course Case Histories," c. November 1938, folder 1, series V.A.1.m, box II, Kinsey Collection, KIA.

53. Before the summer session, both Carroll Christenson and Albert L. Kohlmeier resigned from the course in disagreement with Kinsey's emphasis on the "vulgar" aspects of marriage. An IU law professor, Bernard Gavit, replaced Harper in the legal section of the course, and Kinsey replaced Kohlmeier with a pair of lecturers who would discuss marriage in explicitly religious terms: Rev. W. E. Moore of the local First Christian Church for a Protestant perspective and Father Thomas Kilfoil of St. Charles Borromeo Church for a Catholic perspective. Fowler V. Harper to Alfred C. Kinsey, August 13, 1939, Fowler V. Harper file, Kinsey Correspondence Collection, KIA; W. E. Moore, "Protestant Conception of Marriage," March 20, 1940, folder 14, series V.A.1.e, box 11, Kinsey Collection, KIA; Thomas Kilfoil, "The Ethical Aspects of Marriage," February 19, 1940, folder 5, series V.A.1.e, box II, Kinsey Collection, KIA. For a detailed consideration of Kinsey's interactions with religious figures, see R. Marie Griffith, "The Religious Encounters of Alfred C. Kinsey," *Journal of American History* 95 (September 2008), 349–77.

54. Kinsey to D. and I. W., December 23, 1939, D. and I. W. file, Kinsey Correspondence Collection, KIA; B. C. to Kinsey, early August 1940, B. C. file, ibid. I have made the correspondents of some letters anonymous at the request of the Kinsey Institute.

55. Kinsey to W. O. W., October 9, 1940, W. O. W. file, ibid.

56. Kinsey to W. G. Jr., August 18, 1939, W. G. Jr. file, Kinsey Correspondence Collection, KIA; B. H. to Kinsey, June 21, 1939, B. H. file, ibid.; Kinsey to B. C., August 25, 1939, B. C. file, ibid.; Alfred C. Kinsey to H. E., August 18, 1939, H. E. file, ibid.; Alfred C. Kinsey to O. E., August 18, 1939, O. E. file, ibid.; Kinsey to W. O. W., W. O. W. file, ibid.; Alfred C. Kinsey to G. W. G., May 6, 1939, G. W. G. file, ibid.; Jones, *Alfred C. Kinsey*, 369–74; Kinsey to Voris, c. December 1939, Voris file, ibid.; Kinsey to Voris, July 6, 1939, ibid. Emphasis in original.

57. "Marriage Course to Begin Monday," *Indiana Daily Student*, September 21, 1939, 1; *Indiana Daily Student*, September 22, 1939, 2; "Marriage Course Shows Increase of 45 This Year," *Indiana Daily Student*, September 28, 1939, 1; "Speaking of Courses," *Indiana Daily Student*, September 29, 1939, 4.

58. Compiled student questionnaire data, c. November 1939, folder 10, series V.A.1.i, box II, Kinsey Collection, KIA.

59. Kinsey, "Reproductive Anatomy and Physiology," 13.

60. Kinsey, "Bases of Society," KIA, 12.

61. Edwin H. Sutherland to Wells, January 14, 1940, Wells file, Kinsey Correspondence Collection, KIA; Kinsey, "Bases of Society," 1, KIA; Kinsey, "Reproductive Anatomy and Physiology," July 12, 1938, 13, KIA; Kinsey, "Reproductive Anatomy and Physiology," p. 12, February 21, 1940, folder 6, series V.A.1.e, box II, Kinsey Collection, KIA. Kinsey et al., *Sexual Behavior in the Human Female*, 642–89, esp. 643–44.

62. Kinsey, "Reproductive Anatomy and Physiology," p. 10–11, February 21, 1940; J. Howard Howson, "Emotional Maturity and the Approach to Marriage," in Folsom, *Plan for Marriage*, 60.

63. Alfred C. Kinsey, "Individual Variation," pp. 9, 7, February 28, 1940, folder 8, series V.A.1.e, box II, Kinsey Collection, KIA; Kinsey et al., *Sexual Behavior in the Human Female*.

64. Ibid., 13, 15; Kinsey, Pomeroy, and Martin, *Sexual Behavior in the Human Male*, 563; Kinsey, "Biologic Bases of Society," KIA, 11.

65. Alfred C. Kinsey, "Sex Education," March 6, 1940, pp. 9, 11, folder 10, series V.A.1.e, box II, Kinsey Collection, KIA.

66. Ibid., 12, 14.

67. Kinsey to Wells, August 7, 1940, Wells file, Kinsey Correspondence Collection, KIA; Wells to Kinsey, August 8, 1940, ibid.

68. Kinsey to Voris, c. October 1939, Voris file, Kinsey Correspondence Collection, KIA.

69. Julie Thompson Klein, *Interdisciplinarity: History, Theory, and Practice* (Detroit: Wayne State University Press, 1990), 25; Edgar Anderson to Kinsey, July 16, 1941, Anderson file, Kinsey Correspondence Collection, KIA.

Chapter 4. Ordering Human Sexuality

Epigraph: Hamilton, *Research in Marriage*, xii–xiii.

1. Dickinson himself cites Hamilton and Davis in Dickinson, *Human Sex Anatomy*, 67–70, and 86.
2. Kinsey to Robert Latou Dickinson, July 26, 1946, folder 8a, series II.c, box 1, Robert Latou Dickinson Papers, KIA.
3. Davis, *Factors in the Sex Life of Twenty-Two Hundred Women*, 95–96; Terry, *American Obsession*, 131–35.
4. Davis, *Factors in the Sex Life of Twenty-Two Hundred Women*, 230; Brewer, *Kinsey Interview Kit*, 84–85.
5. Davis, *Factors in the Sex Life of Twenty-Two Hundred Women*, 76–77.
6. Ibid., xv, xviii.
7. Hamilton, *Research in Marriage*, 25, 27, 42–43; Vern L. Bullough, *Science in the Bedroom: A History of Sex Research* (New York: Basic Books, 1994), 118–19; Terry, *American Obsession*, 137–39; Gerhard, *Desiring Revolution*, 38–40.
8. Hamilton, *Research in Marriage*, 16–19; Gorchov, "Sexual Science and Sexual Politics," 142–46, 148–52, esp. 146; Pomeroy, *Dr. Kinsey and the Institute for Sex Research*, 2nd ed., 67–68.
9. Glenn V. Ramsey, interview by James H. Jones, typescript, pp. 6–7, March 15, 1972, CSHM, Indiana University, Bloomington, IN; Kinsey to Raymond Pearl, July 7, 1939, Kinsey folder, Series I, Raymond Pearl Papers, APS, Philadelphia, PA.
10. David Serlin, "Carney Landis and the Psychosexual Landscape of Touch in Mid-Twentieth-Century America," *History of Psychology* 15, no. 3 (2012): 212.
11. Brewer, *Kinsey Interview Kit*, 72–130.
12. Thomas G. Albright, comp. and ed., *The Kinsey Interview Kit: Codebook*, 2nd ed. (Bloomington, IN: Kinsey Institute for Research in Sex, Gender, and Reproduction, 2006), 48, 90, 122.
13. Albright, *Kinsey Interview Kit: Codebook*, 1; Brewer, *Kinsey Interview Kit*, 21.
14. Fernald and Kinsey, *Edible Wild Plants of Eastern North America*.
15. Albright, *Kinsey Interview Kit: Codebook*, 74–84.
16. Pomeroy, *Dr. Kinsey and the Institute for Sex Research*, 129.
17. Kinsey, Pomeroy, and Martin, *Sexual Behavior in the Human Male*, 179.
18. Ibid., 58; Pomeroy, *Dr. Kinsey and the Institute for Sex Research*, 121.
19. Kinsey to Sophia J. Kleegman, March 3, 1948, Kleegman file, Kinsey Correspondence Collection, KIA.
20. Davis, *Factors in the Sex Life of Twenty-Two Hundred Women*, 95; Hamilton, *Research in Marriage*, 25; "N.Y.S.P.I. and H./N.R.C. [New York State Psychiatric Institution and Hospital/National Research Council] Special Investigation," c. 1935, p. 18, folder 5, series III.A, box 1, Carney Landis Papers, KIA; "N.Y.S.P.I. and

H./N.R.C./Form M-W" [New York State Psychiatric Institution and Hospital/National Research Council/Form Married-Women] Special Investigation," c. 1935, p. 4, folder 5, ibid. See Carney Landis et al., *Sex in Development: A Study of the Growth and Development of the Emotional and Sexual Aspects of Personality together with Physiological, Anatomical, and Medical Information on a Group of 153 Normal Women and 142 Female Psychiatric Patients* (New York: Paul B. Hoeber, 1940), 97, 100.

21. Dickinson and Beam, *Single Woman*, xvii; Davis, *Factors in the Sex Life of Twenty-Two Hundred Women*, 76–77.

22. For differing interpretations of Kinsey's use of orgasm as a standard measure, see Ericksen with Steffen, *Kiss and Tell*, 48, 51; Irvine, *Disorders of Desire*, 52.

23. Igo, *Averaged American*, 221.

24. Bullough, *Science in the Bedroom*, 104, 111–12; Max Joseph Exner, *Problems and Principles of Sex Education: A Study of 948 College Men* (New York: Association Press, 1915), 27; David J. Pivar, *Purity and Hygiene: Women, Prostitution, and the "American Plan," 1900–1930* (Westport, CT: Greenwood, 2002), 242.

25. Exner, *Problems and Principles of Sex Education*, quoted in O. L. Harvey, "The Questionnaire as Used in Recent Studies of Human Sexual Behavior," *Journal of Abnormal Psychology* 26, no. 4 (1932): 380.

26. Kinsey, Pomeroy, and Martin, *Sexual Behavior in the Human Male*, 25.

27. Ibid., 36–40, 93–102.

28. Jones, *Alfred C. Kinsey*, 306.

29. "Raymond Pearl Papers," American Philosophical Society, accessed March 1, 2014, http://amphilsoc.org/mole/view?docId=ead/Mss.B.P312-ead.xml.

30. Raymond Pearl, *The Natural History of Population* (New York: Oxford University Press, 1939).

31. Raymond Pearl, Diary, 1938, Raymond Pearl Papers, APS. Pearl did not impress the students enrolled in the fall 1938 marriage course. In the course evaluation summary, one male student referred to Pearl's lecture as "another period to sleep." "Summary of Student Answers," November 1938, folder 5, series V.A.1.i, box II, Kinsey Collection, KIA.

32. Raymond Pearl, *Man the Animal*, ed. Maud M. DeWitt Pearl (Bloomington, IN: Principia Press, 1946), 13, 31 (emphasis in original).

33. Ibid., 34, 31.

34. Ibid., 40.

35. Ibid., 41.

36. Ibid., 43–44. Raymond Pearl, *Introduction to Medical Biometry and Statistics*, 3rd ed. (Philadelphia: W. B. Saunders, 1940).

37. Pearl, *Man the Animal*, 47–128.

38. Christenson, *Alfred C. Kinsey*, 103. I have been unable to find the original letter or the correspondent in the Kinsey Correspondence Collection. Jones, *Alfred C. Kinsey*, 258.

39. Snedecor thanked William Cochran for his input as an Iowa State colleague and fellow statistician on the fourth edition. Cochran led the statistical team that critiqued the *Male* volume and eventually took over as author of subsequent editions of the textbook. George W. Snedecor, *Statistical Methods Applied to Experiments in Agriculture and Biology*, 4th ed. (Ames, IA: Collegiate Press, 1946), ix.

40. Pearl, *Introduction to Medical Biometry and Statistics*, 1, 4.

41. Snedecor, *Statistical Methods Applied to Experiments in Agriculture and Biology*, 1–3, 6, 55–60; Pearl, *Introduction to Medical Biometry and Statistics*, 79–80, 87–88, 130–31. Pearl may also have convinced Kinsey to put his data on punched cards using a punched-card machine. He advocates the use of punched cards for biometric data in *Introduction to Medical Biometry and Statistics*, 122–24.

42. Pearl, *Introduction to Medical Biometry and Statistics*, 9. Emphasis in original.

43. Ibid., 11.

44. Ibid., 12, 13. Emphases in original.

45. Cochran, Mosteller, and Tukey, *Statistical Problems of the Kinsey Report on Sexual Behavior in the Human Male*.

46. Pearl, *Introduction to Medical Biometry and Statistics*, 7.

47. Ibid., 282.

48. Hegarty, *Gentlemen's Disagreement*, 138.

49. Pomeroy, *Dr. Kinsey and the Institute for Sex Research*, 2nd ed., 262.

50. Snedecor, *Statistical Methods Applied to Experiments in Agriculture and Biology*, 70 (emphasis in original), 73, 72. He states that an "objective in sampling is that the estimates shall be as accurate as possible for the time and money spent. It is possible that a nonrandom sampling may accomplish this while a random sample would be less accurate." Ibid., 455.

51. Ibid., 455. Snedecor does not use that phrase himself in his discussion of sampling from a homogenous population, but Kinsey uses the phrase in the *Male* volume to cover what the statistician had called "stratified sampling" earlier in the book.

52. Pearl, *Introduction to Medical Biometry and Statistics*, 93, italics of whole sentence in original removed.

53. The only mention that Paul Robinson makes of the statistics in the *Male* volume is describing Kinsey's sampling as an example of "the principle of distribution." His "principle of distribution" is a muddled account of nonprobability sampling. He also does not analyze the statistical chapter. Robinson, *Modernization of Sex*, 52–53.

54. For the friendship between Pearl and William Morton Wheeler, see generally, William Morton Wheeler Folders 1–7, 1902–22, Series I, Pearl Papers, APS.

55. Kinsey to Raymond Pearl, July 7, 1939, Kinsey folder, Series I, Pearl Papers, APS.

56. Richard von Krafft-Ebing, *Psychopathia Sexualis: With Especial Reference to*

Contrary Sexual Instinct; A Medico-Legal Study (Philadelphia: F. A. Davis, 1892); Sigmund Freud, *Three Essays on the Theory of Sexuality* (New York: Basic Books, 1975).

57. Gathorne-Hardy, *Sex the Measure of All Things*, 186; Arthur L. Norberg, "High-Technology Calculation in the Early 20th Century: Punched Card Machinery in Business and Government," *Technology and Culture* 31, no. 4 (1990): 771; Lars Heide, *Punched-Card Systems and the Early Information Explosion, 1880–1945* (Baltimore: Johns Hopkins University Press, 2009), 5.

58. Norberg, "High-Technology Calculation in the Early 20th Century," 756.

59. Ibid., 768.

60. Wallace Eckert, "Punched Card Methods in Scientific Computation," c. 1939, typescript, pp. 11, 13, 28, folder 16, box 1, Wallace Eckert Papers, Charles Babbage Institute, University of Minnesota Archives, Minneapolis (hereafter cited as CBI).

61. Wallace John Eckert, *Punched Card Methods in Scientific Computation* (New York: Thomas J. Watson Astronomical Computing Bureau, Columbia University, 1940), 10–12; International Business Machines Corporation, "IBM Electric Punched Card Accounting Machines: Principles of Operation; Sorters Types 82, 80 and 75," IBM Form 22–3177–2 (1949; IBM Corporation, New York: 1953), 13–15, "Sorters 80, 82, 75" folder, box 98, series 60, Computer Product Manuals Collection, CBI.

62. Heide, *Punched-Card Systems and the Early Information Explosion*, 6.

63. Brewer, *Kinsey Interview Kit*, 61–62, 18. See also Albright, *Kinsey Interview Kit: Codebook*, and Hegarty, *Gentlemen's Disagreement*, 94.

64. Kinsey et al., *Sexual Behavior in the Human Female*, 61.

65. Brewer, *Kinsey Interview Kit*; Jones, *Alfred C. Kinsey*, 433.

66. Kinsey, Pomeroy, and Martin, *Sexual Behavior in the Human Male*, 71, 72. Kinsey's description of the machine-coding process fit into Bruno Latour's actor-network theory: "Everything that might enhance either the mobility, or the stability, or the combinability of the elements will be welcomed and selected if it accelerates the accumulation cycle." Latour, *Science in Action*, 227.

67. Alfred C. Kinsey, "Studies in Human Sexual Behavior, Progress Report," April 1, 1946, p. 1, *Institute for Sex Research 1938–55: Annual Reports*, KIA.

68. Pomeroy, *Dr. Kinsey and the Institute for Sex Research*, 2nd ed., 87.

69. Kinsey to Ramsey, February 16, 1941, folder 1, Ramsey file, Kinsey Correspondence Collection, KIA; see also Pomeroy, *Dr. Kinsey and the Institute for Sex Research*, 88.

70. Jones, *Alfred C. Kinsey*, 429; Alfred C. Kinsey, "Studies in Human Sex Behavior: Progress Report, Projected Program, 1942–3," March 1942, p. 6, *Institute for Sex Research 1938–55: Annual Reports*, KIA.

71. Jennifer S. Light, "When Computers Were Women," *Technology and Culture* 40, no. 3 (1999): 471; Alfred C. Kinsey, "Studies in Human Sex Behavior: Progress Report," April 1, 1948, p. 7, *Institute for Sex Research 1938–55*, Kinsey Collection,

KIA. In the 1979 volume *Kinsey Data*, Gebhard and Johnson state that for the *Male* and *Female* volumes, "only those portions of our data which we intended to use in publications were punched on cards," though they do not specify what was left out (Paul H. Gebhard and Alan B. Johnson, *The Kinsey Data: Marginal Tabulations of the 1938–1963 Interviews Conducted by the Institute for Sex Research* [Philadelphia: W. B. Saunders, 1979], 37). Two sets of information remained unpunched as of this writing: details relating to reproduction that had been published in Gebhard et al., *Pregnancy, Birth, and Abortion* and "most of the narrative supplementary information on the bottom or reverse of the case history sheet" as "such highly specialized data could wait for a future research project and need not be included in our more basic data storage" (Gebhard and Johnson, *Kinsey Data*, 39). All of the original interview sheets, with or without supplementary information, are stored in the Kinsey Institute Archives.

72. Latour, *Science in Action*, 227, 236.

73. Gebhard and Johnson, *Kinsey Data*, 8.

74. Kinsey, Pomeroy, and Martin, *Sexual Behavior in the Human Male*, 119.

75. Kinsey to Ramsey, March 28, 1941, Ramsey file, Kinsey Correspondence Collection, KIA.

76. Alfred C. Kinsey, "Biologic Bases of Society," p. 11, June 28, 1938, folder 1, series V.A.1.g, box II, Kinsey Collection, KIA.

77. William H. Masters and Virginia E. Johnson, *Human Sexual Response* (Boston: Little, Brown, 1966); Donna J. Drucker, *The Machines of Sex Research: Technology and the Politics of Identity, 1945–1985* (Dordrecht: Springer, 2014).

78. Igo, *Averaged American*, 218.

79. Pomeroy, *Dr. Kinsey and the Institute for Sex Research*, 263–64.

Chapter 5. The Taxonomy and Classification of Human Sexuality

Epigraph: Kinsey, Pomeroy, and Martin, *Sexual Behavior in the Human Male*, 23.

1. Dickinson, *Human Sex Anatomy*, vii. Wardell Pomeroy states that Kinsey was the sole author of the *Male* volume, and so I refer to Kinsey as the sole author. Pomeroy, *Dr. Kinsey and the Institute for Sex Research*, 261–64.

2. Alfred Emerson to Kinsey, January 26, 1948, Emerson file, Kinsey Correspondence Collection, KIA; Kinsey to Emerson, January 28, 1948, ibid.

3. Kinsey, Pomeroy, and Martin, *Sexual Behavior in the Human Male*, 9.

4. Ibid., 88, 92.

5. Ibid., 17.

6. Reumann, *American Sexual Character*, 116. Kinsey states that his unpublished data on African Americans show that "Negro and white patterns for comparative social levels are close if not identical." Kinsey, Pomeroy, and Martin, *Sexual Behavior in the Human Male*, 393.

7. Kinsey, Pomeroy, and Martin, *Sexual Behavior in the Human Male*, 18.

8. Donna J. Drucker, "'A Most Interesting Chapter in the History of Science': Intellectual Responses to Alfred Kinsey's *Sexual Behavior in the Human Male*," *History of the Human Sciences* 25 (February 2012): 75–98.

9. Kinsey, Pomeroy, and Martin, *Sexual Behavior in the Human Male*, 9; Dickinson and Beam, *Thousand Marriages*; Dickinson and Beam, *Single Woman*.

10. Kinsey, Pomeroy, and Martin, *Sexual Behavior in the Human Male*, 9.

11. Ibid. (my emphasis); Igo, *Averaged American*, 205.

12. Kinsey, Pomeroy, and Martin, *Sexual Behavior in the Human Male*, 27; Exner, *Problems and Principles of Sex Education*.

13. Kinsey, Pomeroy, and Martin, *Sexual Behavior in the Human Male*, 10; Hamilton, *Research in Marriage*; Lilburn Merrill, "A Summary of Findings in a Study of Sexualism among a Group of One Hundred Delinquent Boys," *Journal of Delinquency* 3 (November 1918): 255–67; Carney Landis et al., *Sex in Development*.

14. Kinsey, Pomeroy, and Martin, *Sexual Behavior in the Human Male*, 10, 183, 185.

15. Ibid., 158. Hamilton, *Research in Marriage*, 25; "N.Y.S.P.I. and H./N.R.C. Special Investigation," 18. Though there is much scientific and public debate about physiological similarities and differences between men's and women's orgasms, Elisabeth A. Lloyd argues that they are physiologically similar in *The Case of the Female Orgasm: Bias in the Science of Evolution* (Cambridge, MA: Harvard University Press, 2005), 109.

16. Kinsey, Pomeroy, and Martin, *Sexual Behavior in the Human Male*, 159, 193; Robinson, *Modernization of Sex*, 57–58.

17. Kinsey, Pomeroy, and Martin, *Sexual Behavior in the Human Male*, 29, 34.

18. Ibid., 23, 29; Glenn V. Ramsey, "Sex Information of Younger Boys," *American Journal of Orthopsychiatry* 13 (April 1943): 347–52; Glenn V. Ramsey, "Sexual Development of Boys," *American Journal of Psychology* 56 (April 1943): 217–34; Kenneth Martin Peterson, "Early Sex Information and Its Influence on Later Sex Concepts" (MA thesis, University of Colorado, Boulder, 1938).

19. Pearl, *Introduction to Medical Biometry and Statistics*, 9; Kinsey, Pomeroy, and Martin, *Sexual Behavior in the Human Male*, 20 (emphasis in original).

20. Kinsey, Pomeroy, and Martin, *Sexual Behavior in the Human Male*, 82.

21. Ibid., 71.

22. Brewer, *Kinsey Interview Kit*, 15–16.

23. Kinsey, Pomeroy, and Martin, *Sexual Behavior in the Human Male*, 82 (my emphasis), 92; Snedecor, *Statistical Methods*.

24. Kinsey, Pomeroy, and Martin, *Sexual Behavior in the Human Male*, 89; Pearl, *Man the Animal*, 43.

25. Kinsey, Pomeroy, and Martin, *Sexual Behavior in the Human Male*, 89, 88, 105, 93; Lewis M. Terman, "Kinsey's 'Sexual Behavior in the Human Male': Some Comments and Criticisms," *Psychological Bulletin* 45, no. 5 (1948): 444.

26. Kinsey, Pomeroy, and Martin, *Sexual Behavior in the Human Male*, 102, 104, 109, 128–29, 135, 142, 153; Igo, *Averaged American*, 209–10, 217.

27. Kinsey, Pomeroy, and Martin, *Sexual Behavior in the Human Male*, 329, 335.

28. Ibid., 36.

29. Ibid., 327, 330.

30. Paul Robinson argues that Kinsey identified two levels according to wealth (rich and poor) when in fact Kinsey's two social level scales are based on occupational class and education. Robinson, *Modernization of Sex*, 92–99.

31. W. Lloyd Warner and Paul S. Lunt, *Social Life of a Modern Community*; Sorokin, *Social Mobility*; F. Stuart Chapin, *The Measurement of Social Status by the Use of the Social Status Scale 1933* (Minneapolis: University of Minnesota Press, 1933).

32. Warner and Lunt, *Social Life of a Modern Community*, 82, 88. They noneuphonically called their six classes lower-lower, upper-lower, lower-middle, upper-middle, lower-upper, and upper-upper.

33. Kinsey, Pomeroy, and Martin, *Sexual Behavior in the Human Male*, 331.

34. Ibid., 331, 330.

35. Ibid., 339, 347, 351, 375–81.

36. Ibid., 630, 652.

37. Ibid., 418; Warner and Lunt, *Social Life of a Modern Community*.

38. Kinsey, Pomeroy, and Martin, *Sexual Behavior in the Human Male*, 395, 397, 418, 427–33.

39. Ibid., 436, 437, 445; Warner and Lunt, *Social Life of a Modern Community*, 102, 104.

40. Kinsey, Pomeroy, and Martin, *Sexual Behavior in the Human Male*, 442.

41. Ibid., 414–15, 417; Sorokin, *Social Mobility*, 170.

42. Kinsey, Pomeroy, and Martin, *Sexual Behavior in the Human Male*, 417–39, esp. 438–39; Warner and Lunt, *Social Life of a Modern Community*, 91.

43. Kinsey, Pomeroy, and Martin, *Sexual Behavior in the Human Male*, 440, 332.

44. Ibid., 202.

45. Ibid., 681.

46. Ibid.

47. Ibid.

48. Hegarty, *Gentlemen's Disagreement*, 46.

49. Kinsey, Pomeroy, and Martin, *Sexual Behavior in the Human Male*, 397.

50. Ibid., 372–73.

51. Ibid., 544.

52. Ibid., 299–300.

53. Ibid., 300.

54. Ibid., 545.

55. Ibid.

56. Bullough, *Science in the Bedroom*, 176; Lewis M. Terman and Catharine C. Miles, *Sex and Personality: Studies in Masculinity and Femininity* (New York:

McGraw-Hill, 1936); Kinsey to Lewis M. Terman, August 23, 1941, Terman file, Kinsey Correspondence Collection, KIA.

57. Kinsey to Glenn V. Ramsey, September 20, 1940, folder 1, Ramsey file, Kinsey Correspondence Collection, KIA. The diagram is not extant. Glenn V. Ramsey, "Factors in the Sex Life of 291 Boys" (EdD diss., Indiana University, Bloomington, 1941), 220–25. For a more detailed account of Ramsey's work and its impact on Kinsey's thinking, see Donna J. Drucker, "Beginning the Kinsey Reports: Glenn Ramsey's Sex Research in Peoria, 1938–1941," *Journal of Illinois History* 10 (Winter 2007): 271–88.

58. Kinsey, "Criteria for a Hormonal Explanation of the Homosexual," 424, 425; Stephanie Hope Kenen, "Who Counts When You're Counting Homosexuals? Hormones and Homosexuality in Mid-Twentieth-Century America," in *Science and Homosexualities*, ed. Vernon A. Rosario (New York: Routledge, 1997), 197–218.

59. Kinsey to Ramsey, February 16, 1941, folder 1, Ramsey file, Kinsey Correspondence Collection, KIA.

60. Henry L. Minton, *Departing from Deviance: A History of Homosexual Rights and Emancipatory Science in America* (Chicago: University of Chicago Press, 2002), 164; Kinsey to Ramsey, September 20, 1940, KIA; Kinsey, "Criteria for a Hormonal Explanation of the Homosexual," 424–28.

61. James Gilbert, *Men in the Middle: Searching for Masculinity in the 1950s* (Chicago: University of Chicago Press, 2005), 86; Stephanie Hope Kenen, "Scientific Studies of Human Sexual Difference in Interwar America" (PhD diss., University of California, Berkeley, 1998), 246. See also Minton, *Departing from Deviance*, 168; and Garton, *Histories of Sexuality*, 205.

62. Kinsey, Pomeroy, and Martin, *Sexual Behavior in the Human Male*, 647.

63. Kinsey et al., *Sexual Behavior in the Human Female*, 471, 469.

64. Brewer, *Kinsey Interview Kit*, 129; Donna J. Drucker, "Male Sexuality and Alfred Kinsey's 0–6 Scale: Toward 'A Sound Understanding of the Realities of Sex,'" *Journal of Homosexuality* 57, no. 9 (2010): 1105–23.

65. Hegarty, *Gentlemen's Disagreement*, 76.

66. Kinsey, Pomeroy, and Martin, *Sexual Behavior in the Human Male*, 639, 641.

67. Ibid., 641.

68. Ibid.

69. Ibid.; see also Hegarty, *Gentlemen's Disagreement*, 76.

70. Kinsey, Pomeroy, and Martin, *Sexual Behavior in the Human Male*, 656.

71. K. A. Cuordileone, *Manhood and American Political Culture in the Cold War* (New York: Routledge, 2004); David K. Johnson, *The Lavender Scare: The Cold War Persecutions of Gays and Lesbians in the Federal Government* (Chicago: University of Chicago Press, 2004).

72. Kinsey, Pomeroy, and Martin, *Sexual Behavior in the Human Male*, 474–75.

73. Ibid., 616.

74. For examples of how Internet users discuss the scale in the present, see Donna J. Drucker, "Marking Sexuality from 0–6: The Kinsey Scale in Online Culture," *Sexuality and Culture* 16 (September 2012): 241–62; see also Donald P. McWhirter, Stephanie A. Sanders, and June M. Reinisch, eds., *Homosexuality/Heterosexuality: Concepts of Sexual Orientation* (New York: Oxford University Press, 1990).

75. Kinsey, Pomeroy, and Martin, *Sexual Behavior in the Human Male*, 327.

Chapter 6. The Boundaries of Sexual Categorization

Epigraph: Kinsey et al., *Sexual Behavior in the Human Female*, 3.

1. Kinsey Institute for Research in Sex, Gender, and Reproduction, "Media Responses to the Kinsey Report," accessed March 2, 2014, http://www.kinseyinstitute.org/library/mediaresponses.html.

2. Pomeroy, *Dr. Kinsey and the Institute for Sex Research*, 266; Ron Jackson Suresha, "'Properly Placed before the Public': Publication and Translation of the Kinsey Reports," *Journal of Bisexuality* 8, nos. 3–4 (2008): 203–28.

3. Drucker, "Most Interesting Chapter," 75–98.

4. Kinsey, Pomeroy, and Martin, *Sexual Behavior in the Human Male*, 7.

5. Kinsey et al., *Sexual Behavior in the Human Female*, 22; Pomeroy, *Dr. Kinsey and the Institute for Sex Research*, 328.

6. Pomeroy, *Dr. Kinsey and the Institute for Sex Research*, 343.

7. Kinsey et al., *Sexual Behavior in the Human Female*, 3; Kinsey, Pomeroy, and Martin, *Sexual Behavior in the Human Male*, 3.

8. Kinsey et al., *Sexual Behavior in the Human Female*, 83.

9. The *Female* volume, as Wardell Pomeroy described it, was more of a group project than was the *Male* volume. However, Kinsey still authored all of the text, and I refer to him as the sole author. Pomeroy, *Dr. Kinsey and the Institute for Sex Research*, 2nd ed., 330–31, 339; see also Donna J. Drucker, "Creating the Kinsey Reports: Intellectual and Methodological Influences on Alfred Kinsey's Sex Research, 1919–1953" (PhD diss., Indiana University, 2008), 177–78.

10. Jones, *Alfred C. Kinsey*, 635–65; Gathorne-Hardy, *Sex the Measure of All Things*, 374–75; Dorothy Craig Collins interview by James H. Jones, typescript, pp. 4–5, December 9, 1971, CSHM.

11. Kinsey et al., *Sexual Behavior in the Human Female*, 26.

12. For a detailed account of the ASA team's visits to Bloomington, see Drucker, "Creating the Kinsey Reports," 160–66.

13. Kinsey et al., *Sexual Behavior in the Human Female*, 54.

14. Kinsey to William I. Cochran, January 12, 1951, Kinsey 1951 folder, Series VI, John W. Tukey Papers, APS; see also Cochran to Kinsey, February 1, 1951, ibid.

15. Kinsey et al., *Sexual Behavior in the Human Female*, 22.

16. For examples of such stereotyping, see Regina Kunzel, *Fallen Women, Problem Girls: Unmarried Mothers and the Professionalization of Social Work, 1890–1945* (New Haven, CT: Yale University Press, 1995); and Ruth Feldstein, *Motherhood in Black and White: Race and Sex in American Liberalism, 1930–1965* (Ithaca: Cornell University Press, 2000).

17. Kinsey et al., *Sexual Behavior in the Human Female*, 7.

18. Ibid., 330, 487, 437; Pomeroy, *Dr. Kinsey and the Institute for Sex Research*, 343.

19. The ISR's book on incarcerated men and women did not come out until 1965, and it focused only on sex offenders. Paul H. Gebhard et al., *Sex Offenders: An Analysis of Types* (New York: Harper & Row, 1965).

20. Kinsey to Cochran, January 12, 1951.

21. Kinsey et al., *Sexual Behavior in the Human Female*, 55, 150, 202, 242, 296, 356, 422, 461.

22. Karl S. Lashley to Kinsey, August 28, 1952, Lashley file, Kinsey Correspondence Collection, KIA; Lashley to Kinsey, February 23, 1953, ibid.

23. Kinsey to Lashley, July 28, 1952, ibid.

24. Kinsey et al., *Sexual Behavior in the Human Female*, 192.

25. Ibid., 132.

26. Ibid., 164.

27. Ibid., 253, 258.

28. Beach was a psychobiologist who studied sexual behavior in mammals whom Kinsey had first met at a 1943 CRPS meeting, and who quickly became Kinsey's collaborator, colleague, and friend. Frank A. Beach, interview by James H. Jones, typescript, p. 5, August 20, 1971, CSHM; Levens, "Sex, Neurosis, and Animal Behavior," 278–79.

29. Kinsey et al., *Sexual Behavior in the Human Female*, 27–28.

30. Ibid., 135–36, 196, 232, 284; Fausto-Sterling, *Sexing the Body*, 225.

31. Kinsey et al., *Sexual Behavior in the Human Female*, 134–36, 195–96, 228–30, 282–84, 346–47, 410–15, 448–52, 503–4.

32. Ibid., 134n7, 135n7, 135n8. Frank A. Beach, "A Cross-Species Survey of Mammalian Sexual Behavior," in *Psychosexual Development in Health and Disease*, ed. Paul H. Hoch and Joseph Zubin (New York: Grune & Stratton, 1949), 63; Havelock Ellis, *Studies in the Psychology of Sex*, vol. 1, 3rd ed. (Philadelphia: F. A. Davis, 1910), 165; Albert R. Shadle, Marilyn S. Smelzer, and Margery M. Metz, "The Sex Reactions of Porcupines (*Erethizon d. dorsatum*) before and after Copulation," *Journal of Mammalogy* 27 (May 1946): 116–21; Clellan S. Ford and Frank A. Beach, *Patterns of Sexual Behavior* (1951; New York: Harper Colophon, 1972), 166; Malinowski, *Sexual Life of Savages in North-Western Melanesia*, 55–59. For communication between Kinsey and Shadle, see Kinsey to Shadle, January 26, 1948, Shadle file, Kinsey Correspondence Collection, KIA; Kinsey to Shadle, August 11, 1950, ibid.; Kinsey to Shadle, July 26, 1951, ibid.; Kinsey to Shadle, January 5, 1953, ibid.

33. Ford and Beach, *Patterns of Sexual Behavior*, 158.

34. Kinsey et al., *Sexual Behavior in the Human Female*, 136.

35. Ibid., 227.

36. Shadle, Smelzer, and Metz, "Sex Reactions of Porcupines"; Ford and Beach, *Patterns of Sexual Behavior*, 42, 58, 66, 192–93; Frank A. Beach, *Hormones and Behavior: A Survey of Interrelationships between Endocrine Secretions and Patterns of Overt Response* (New York: Paul B. Hoeber, 1948).

37. Kinsey et al., *Sexual Behavior in the Human Female*, 228–31, esp. 230.

38. Clellan S. Ford, *A Comparative Study of Human Reproduction* (New Haven, CT: Yale University Press, 1945), 100; Jules Henry, "The Social Function of Child Sexuality in Pilaga Indian Culture," in Hoch and Zubin, *Psychosexual Development in Health and Disease*, 94–98; Malinowski, *Sexual Life of Savages in North-Western Melanesia*, 340, 376; Ford and Beach, *Patterns of Sexual Behavior*, 183. Those works are cited in Kinsey et al., *Sexual Behavior in the Human Female*, 284n5, 108n8, 136n8, and 284n4, respectively.

39. Ford and Beach, *Patterns of Sexual Behavior*, 75–77, 211–20, 113, 116; see also Ford, *Comparative Study of Human Reproduction*, 12–13, 28–30, 93.

40. Kinsey et al., *Sexual Behavior in the Human Female*, 8.

41. Ibid., 411–13, esp. 411n5 and 412; Hamilton, "Study of Sexual Tendencies in Monkeys and Baboons," 303; E. J. Kempf, "The Social and Sexual Behavior of Infra-human Primates with Some Comparable Facts in Human Behavior," *Psychoanalytic Review* 4 (1917): 143.

42. Kinsey et al., *Sexual Behavior in the Human Female*, 433.

43. Beach, "Cross-Species Survey of Mammalian Sexual Behavior," 63–64; Beach, *Hormones and Behavior*, 36, 139; Ellis, *Studies in the Psychology of Sex*, vol. 3, 165; Ford and Beach, *Patterns of Sexual Behavior*, 139; Beach, "Hormones and Mating Behavior in Vertebrates," in *Recent Progress in Hormone Research*, ed. Gregory Pincus, vol. 1 (New York: Academic Press, 1947), 41; Michael Pettit, "The Queer Life of a Lab Rat," *History of Psychology* 15, no. 3 (2012): 217–27.

44. Kinsey et al., *Sexual Behavior in the Human Female*, 450n6; Beach, "Cross-Species Survey of Mammalian Sexual Behavior," 64–65; Ford and Beach, *Patterns of Sexual Behavior*, 136, 139.

45. Beach, "Cross-Species Survey of Mammalian Sexual Behavior," 64–65; Ford and Beach, *Patterns of Sexual Behavior*, 141–42; Beach, "Hormones and Mating Behavior in Vertebrates," 66–68.

46. Kinsey et al., *Sexual Behavior in the Human Female*, 449n2, 450n6.

47. Ford and Beach, *Patterns of Sexual Behavior*, 133, 143.

48. Kinsey et al., *Sexual Behavior in the Human Female*, 450n6, see also 449n2, 452n9; Ford and Beach, *Patterns of Sexual Behavior*, 133, 143; Davis, *Factors in the Sex Life of Twenty-Two Hundred Women*, 161, 97; and Beach, "Cross-Species Survey of Mammalian Sexual Behavior," 63–64.

49. Pomeroy, *Dr. Kinsey and the Institute for Sex Research*, 181–84; Gathorne-Hardy, *Sex the Measure of All Things*, 307, 333–36. Samuel Steward's experience of participating in a sadomasochistic attic film is described in Justin Spring, *Secret Historian: The Life and Times of Samuel Steward, Professor, Tattoo Artist, and Sexual Renegade* (New York: Farrar, Straus, and Giroux, 2010), 139–41, and fictional renditions of the films are depicted in the movie *Kinsey* (2004) and the novel *Inner Circle*, by T. C. Boyle.

50. Kinsey et al., *Sexual Behavior in the Human Female*, 3.

51. Levens, "Sex, Neurosis, and Animal Behavior," 227.

52. Frank A. Beach, "A Review of Physiological and Psychological Studies of Sexual Behavior in Mammals," *Physiological Reviews* 27 (June 1947): 246.

53. Kinsey et al., *Sexual Behavior in the Human Female*, 229n3, see also ibid., 586–88.

54. Ford and Beach, *Patterns of Sexual Behavior*, 58–59; see also Albert R. Shadle, "Copulation in the Porcupine," *Journal of Wildlife Management* 10 (April 1946): 159–62.

55. Kinsey et al., *Sexual Behavior in the Human Female*, 584; Judith A. Allen, "Kinsey's Women: Sexed Bodies, Heterosexualities, and Abortion, 1920–1960," typescript, Department of Gender Studies, Indiana University, 2002, ch. 2, pp. 16–18; Sophia Kleegman to Kinsey, November 4, 1950, Kleegman file, Kinsey Correspondence Collection, KIA; Gorchov, "Sexual Science and Sexual Politics," 192–93.

56. Kinsey et al., *Sexual Behavior in the Human Female*, 582n12, 583n13.

57. Kleegman to Kinsey, December 21, 1950, Kleegman file, KIA; Kleegman to Kinsey, May 10, 1951, ibid.; Kinsey to Francis J. Hector, June 15, 1951, Hector file, Kinsey Correspondence Collection, KIA; Kinsey to Carl O. Hartman, Dec. 11, 1951, Hartman file, Kinsey Correspondence Collection, KIA; Kinsey et al., *Sexual Behavior in the Human Female*, x.

58. Kinsey et al., *Sexual Behavior in the Human Female*, 626.

59. Eduard E. Hitschmann and Edmund Bergler, *Frigidity in Women: Its Characteristics and Treatment*, trans. Polly Leeds Weil (Washington, DC: Nervous and Mental Disease Publishing Co., 1936).

60. Kinsey et al., *Sexual Behavior in the Human Female*, 595–99; Ford and Beach, *Patterns of Sexual Behavior*, 245–58.

61. Kinsey, Pomeroy, and Martin, *Sexual Behavior in the Human Male*, 579–80; Kinsey et al., *Sexual Behavior in the Human Female*, 626.

62. Kinsey et al., *Sexual Behavior in the Human Female*, 468, see also 173–75, 487–89.

63. Ford and Beach, *Patterns of Sexual Behavior*, 241; Levens, "Sex, Neurosis, and Animal Behavior," 247.

64. Kinsey et al., *Sexual Behavior in the Human Female*, 669.

65. Ibid., 650.

66. Kinsey, Pomeroy, and Martin, *Sexual Behavior in the Human Male*, 650, 670, 687; Kinsey et al., *Sexual Behavior in the Human Female*, 475.

67. Fausto-Sterling, *Sexing the Body*, 179.

68. Kinsey et al., *Sexual Behavior in the Human Female*, 729.

69. Ibid., 745. Human sterilization may have had particular resonance for Kinsey given Indiana's long history of eugenic law, beginning with the sterilization law of 1907. Thurman B. Rice, one of Kinsey's local adversaries, was an advocate of eugenics and may have drawn Kinsey's attention to Indiana's laws. See Alexandra Minna Stern, "'We Cannot Make a Silk Purse out of a Sow's Ear': Eugenics in the Hoosier Heartland," *Indiana Magazine of History* 103 (March 2007): 3–9, 37.

70. Robinson, *Modernization of Sex*, 113; Kinsey et al., *Sexual Behavior in the Human Female*, 721n7, 724, 726–27, 731–34, 737–38, 743, 745, 747–51, 753n65, 753n66, 758; Beach, *Hormones and Behavior*, 9–10, 12–17, 20–28, 34, 36, 38–41, 45, 61–62, 68, 203–4, 206–7; Ford and Beach, *Patterns of Sexual Behavior*, 41–42, 142, 167–70, 221–26, 227, 229–32; *Textbook of Endocrinology*, ed. Robert H. Williams (Philadelphia: W. B. Saunders, 1950), 7, 11–76, 115, 122, 170–75; Levens, "Sex, Neurosis, and Animal Behavior," 279.

71. Kinsey et al., *Sexual Behavior in the Human Female*, 761.

72. Fausto-Sterling, *Sexing the Body*, 210; Levens, "Sex, Neurosis, and Animal Behavior," 244–47; Ford and Beach, *Patterns of Sexual Behavior*, 241.

73. Kinsey et al., *Sexual Behavior in the Human Female*, 688–89.

74. Ibid., 690, 712.

75. Ibid., 669–72.

76. He wrote in answer to one inquiry, "We have primarily been interested in sexual behavior, of which reproduction is an incidental part, and we have not attempted to study reproduction *per se*." Kinsey to P. P. Bliss, June 8, 1956, Bliss file, Kinsey Correspondence Collection, KIA. The literature on gender in the postwar United States is vast; for an introduction, see *Not June Cleaver: Women and Gender in Postwar America, 1945–1960*, ed. Joanne Meyerowitz (Philadelphia: Temple University Press, 1994).

77. Robinson, *Modernization of Sex*, 114–15.

78. Kinsey et al., *Sexual Behavior in the Human Female*, 537, 538. See also 146, 230.

79. Ibid., 451, 646, 679–81. For Kinsey's contacts with Harry Benjamin and transsexuals and transgender persons, see Joanne Meyerowitz, *How Sex Changed: A History of Transsexuality in the United States* (Cambridge, MA: Harvard University Press, 2004), 46–48, 108, 154–55, 171, 175, 213; on Kinsey and transsexuality generally, see Meyerowitz, "Sex Research at the Borders of Gender: Transvestites, Transsexuals, and Alfred C. Kinsey," *Bulletin of the History of Medicine* 75 (Spring 2001): 72–90.

Conclusion

Epigraph: Edmund E. Jeffers to Kinsey, January 31, 1948, Jeffers file, Alfred C. Kinsey Correspondence Collection, KIA.

1. Kinsey et al., *Sexual Behavior in the Human Female*, 9.
2. Irvine, "From Difference to Sameness," 23.
3. Kinsey et al., *Sexual Behavior in the Human Female*, 712.
4. Bullough, *Science in the Bedroom*, 210; John Money, "Hermaphroditism, Gender and Precocity in Hyperadrenocorticism: Psychologic Findings," *Bulletin of the Johns Hopkins Hospital* 96 (1955): 253–264.
5. Kinsey et al., *Sexual Behavior in the Human Female*, 666.
6. Ibid., 192, 712.
7. For Internet use of the 0–6 scale, see Drucker, "Marking Sexuality from 0–6," 241–62. On the positive and negative implications of the 10 percent figure, see Drucker, "Male Sexuality and Alfred Kinsey's 0–6 Scale," 1105–23.
8. Igo, *Averaged American*, 299.
9. Kinsey to Yerkes, [fall 1943], quoted in Pomeroy, *Dr. Kinsey and the Institute of Sex Research*, 172.
10. Irvine, "From Difference to Sameness," 11–12.
11. Kinsey to Albert P. Blair, July 5, 1956, Blair file, Kinsey Correspondence Collection, KIA. Blair, along with Herman Spieth, Robert Bugbee, Robert Kroc, and Osmond Breeland—all former colleagues or graduate students of Kinsey's—were among the presenters. See "Indiana University Jordan Hall Dedication Program, June 6–9, 1956," accessed March 2, 2014, http://www.bio.indiana.edu/about/history/JH_dedication_1956.pdf. For excerpts from Kinsey's other letters around the same time expressing his optimism about recovery, see Pomeroy, *Dr. Kinsey and the Institute for Sex Research*, 437–39. Kinsey died August 25, 1956, at the age of sixty-two.
12. Jones, *Alfred C. Kinsey*, 627, 722–24.
13. Edwards et al., "Historical Perspectives on the Circulation of Information," 1421.
14. Shera, *Libraries and the Organization of Knowledge*, 86.

Bibliography

Chronological List of Publications by Alfred C. Kinsey

Kinsey, Alfred C. "New Species and Synonymy of American Cynipidae." *Bulletin of the American Museum of Natural History* 42 (December 20, 1920): 293–317.

———. "Life Histories of American Cynipidae." *Bulletin of the American Museum of Natural History* 42 (December 20, 1920): 319–57.

———. "Phylogeny of Cynipid Genera and Biological Characteristics." *Bulletin of the American Museum of Natural History* 42 (December 20, 1920): 357a–c, 358–402.

———. "New Pacific Coast Cynipidae (Hymenoptera)." *Bulletin of the American Museum of Natural History* 46 (April 24, 1922): 279–95.

———. "Studies of Some New and Described Cynipidae (Hymenoptera)." *Indiana University Studies* 9 (June 1922): 3–141.

———. "The Gall Wasp Genus *Neuroterus* (Hymenoptera)." *Indiana University Studies* 10, no. 58 (June 1923): 3–147.

———. *An Introduction to Biology*. Philadelphia: J. B. Lippincott, 1926.

———. "Biologic Sciences in Our High Schools." *Proceedings of the Indiana Academy of Science* 35 (1925): 63–66.

———. *Field and Laboratory Manual in Biology*. Philadelphia: J. B. Lippincott, 1927.

———. "The Gall Wasp Genus *Cynips*: A Study in the Origin of Species." *Indiana University Studies* 84–86 (June, September, December 1929).

———. "The Content of the Biology Course." *School Science and Mathematics* 30, no. 4 (April 1930): 375–84.

———. "Cynipoidea from Oceania." *Contributions from the Zoological Laboratories of Indiana University*. 268 (1930[?]): 1–8.

———. *New Introduction to Biology*. Chicago: J. B. Lippincott, 1933.

———. *Workbook in Biology*. Chicago: J. B. Lippincott, 1934.

———. *The Origin of Higher Categories in* Cynips. Bloomington: Indiana University Publications, 1936.

———. "An Evolutionary Analysis of Insular and Continental Species." *Proceedings of the National Academy of Sciences of the United States of America* 23 (January 15, 1937): 5–11.

———. "Supra-specific Variation in Nature and in Classification from the View-Point of Zoology." *American Naturalist* 71 (May–June 1937): 206–22.

———. *Methods in Biology*. Chicago: J. B. Lippincott, 1937.

Taylor, L. W., R. W. Taylor, and Alfred C. Kinsey. "An Invitation to Teachers of Science." *Science*, n.s., 87, no. 2267 (June 10, 1938): 528.

———. "New Figitidae from the Marquesas Islands." *Contributions from the Zoological Laboratories of Indiana University*. 273 [1938?]: 193–97.
———. *New Introduction to Biology*. 2nd ed. Chicago: J. B. Lippincott, 1938.
———. "New Mexican Gall Wasps (Hymenoptera, Cynipidae)." *Contributions from the Zoological Laboratories of Indiana University*. 272 [1938?]: 261–77.
———. "Criteria for a Hormonal Explanation of the Homosexual." *Journal of Clinical Endocrinology* 1 (May 1941): 424–28.
———. "Isolating Mechanisms in Gall Wasps." *Biological Symposia* 6 (1942): 251–69.
———. "Seasonal Factors in Gall Wasp Distribution." *Biological Symposia* 6 (1942): 167–87.
Fernald, Merritt Lyndon, and Alfred Charles Kinsey. *Edible Wild Plants of Eastern North America*. Cornwall-on-Hudson, NY: Idlewild Press, 1943.
Kinsey, Alfred C., Wardell B. Pomeroy, and Clyde E. Martin. *Sexual Behavior in the Human Male*. Philadelphia: W. B. Saunders, 1948.
Kinsey, Alfred C., Wardell B. Pomeroy, and Clyde E. Martin. "Concepts of Normality and Abnormality in Sexual Behavior." In Hoch and Zubin, *Psychosexual Development in Health and Disease*, 11–32.
Kinsey, Alfred C., Wardell B. Pomeroy, Clyde E. Martin, and Paul H. Gebhard. *Sexual Behavior in the Human Female*. Philadelphia: W. B. Saunders, 1953.

Manuscript Collections

American Museum of Natural History Library, New York, New York
  Central Archives
  Alfred C. Kinsey Entomological Papers [1917?–1941]
American Philosophical Society, Philadelphia, Pennsylvania
  American Eugenics Society Papers
  William Ernest Castle Papers
  George W. Corner Papers
  Charles B. Davenport Papers
  L. C. Dunn Papers
  Raymond Pearl Papers
  Oscar Riddle Papers
  Anne Roe Papers
  George Gaylord Simpson Papers
  Society for the Study of Evolution Papers
  John W. Tukey Papers
  Sewall Wright Papers
Archives of the Arnold Arboretum of Harvard University, Jamaica Plain, Massachusetts
  Bussey Institution Records, 1872–2007
  Elmer Drew Merrill Papers

Bibliography

Karl Sax (1892–1973) Papers, 1938–2001
Archives of the Gray Herbarium, Harvard University, Cambridge, Massachusetts
  Administrative Correspondence Files
  Merritt Lyndon Fernald Papers
Bancroft Library, University of California, Berkeley, California
  Richard Benedict Goldschmidt Papers, BANC MSS 72/241 z
Bentley Historical Library, University of Michigan, Ann Arbor, Michigan
  Marston Bates Papers
  Marriage Relations Course Files, Michigan Historical Collection
Center for the Study of History and Memory, Indiana University, Bloomington, Indiana
  History: Kinsey Institute for Sex Research (1971–1972), Collection #55
   1. Frank A. Beach
   2. Robert E. Bugbee
   3. Dorothy Craig Collins
   4. Frank K. Edmondson
   5. Margaret Edmondson
   6. Clara McMillen (Mrs. Alfred C.) Kinsey
   7. Clyde E. Martin
   8. Kate Hevner Mueller
   9. Glenn V. Ramsey
   10. Edith B. Schuman
   11. Herman B Wells
Charles Babbage Institute, University of Minnesota Archives, Minneapolis
  Computer Product Manuals Collection
  Wallace Eckert Papers
Ernst Mayr Library, Museum of Comparative Zoology, Harvard University, Cambridge, Massachusetts
  Annual Report of the Director of the Museum of Comparative Zoology at Harvard College to the President and Fellows of Harvard College for 1922–1923.
Indiana University Archives, Bloomington, Indiana
  Board of Trustees Minutes, vol. II (1938)
  Alfred C. Kinsey Papers
  Kinsey-Marriage Course File, Herman B Wells Papers
Harvard University Archives, Cambridge, Massachusetts
  General Information by and about the Bussey Institution
  Papers of William Morton Wheeler
Indiana University–Purdue University Indianapolis Archives, Indianapolis, Indiana
  Dean W. D. Gatch Files
  Indiana University School of Medicine, Dean's Office Files

Kinsey Institute Archives, Bloomington, Indiana
Dorothy Dunbar Bromley Papers
Alice Withrow Field Papers
Robert Latou Dickinson Papers
Alfred C. Kinsey Collection
    "Studies of Gall Wasps (Cynipidae, Hymenoptera)" (ScD diss., Harvard University, 1919) (Box I, Series I.D.1)
    "The Right to Do Research" (includes "Last Statement") (Box I, Series I.D)
    Biology Conferences and Biology Teaching (Box I, Series I.E.1)
    Science Education (Box I, Series I.E.2)
    Notes from Evolution Class, 1951 (Box 1, Series 1.E.3)
    Memoir Series [letters to Paul H. Gebhard] (Box I, Series I.F.1)
    Oral History Interviews (Box I, Series I.F.4)
    Lecture Notes from Bussey Institute Courses (Box I, Series I.I.2)
    Research Reports (Box I, Series II.A.1)
    Biology-Related Periodical Articles (Box I, Series II.D)
    Research Plans (Box I, Series II.B.1)
    Entomological Drawings (Box II, Series III.A)
    Publications (Box II, Series IV.C)
    Marriage Course Lectures [Summer 1938] (Box II, Series V.A.1.b)
    Marriage Course Lectures [Summer 1939] (Box II, Series V.A.1.d)
    Marriage Course Lectures [Spring 1940] (Box II, Series V.A.1.e)
    Medical School Marriage Course Lectures [Fall 1940] (Box II, Series V.A.1.f)
    Marriage Course Lectures [undated] (Box II, Series V.A.1.g)
    Marriage Course Student Evaluations (Box II, Series V.A.1.i)
    Letters to Marriage Course Staff, IU President, and Trustees (Box II, Series V.A.1.j)
    Reading List for Marriage Course Staff (Box II, Series V.A.1.k)
    Indiana University Marriage Course Case Histories (Box II, Series V.A.1.m)
    Evolution Seminar (Box II, Series V.A.2.a)
    Species Seminar (Box II, Series V.A.4)
    Zoology Seminar (Box II, Series V.A.5)
    Teaching Biology and Methods in Biology (Box II, Series V.A.6)
    Evolution Series (Box II, Series V.A.7)
Alfred C. Kinsey Correspondence Collection [1920–1959]
Some names are anonymized at the request of the Kinsey Institute Archives.
    1. American Association of Marriage Counselors
    2. American Museum of Natural History [Ernst Mayr]
    3. Edgar Anderson

4. Marston Bates
5. Albert P. Blair
6. P. P. Bliss
7. Bloomington Hospital School of Nursing
8. Osmond P. Breland
9. C. T. Brues
10. J. M. C.
11. B. C.
12. M. A. C.
13. Mrs. R. L. C.
14. B. C.
15. Alfred Emerson
16. H. E.
17. O. E.
18. W. G. Jr.
19. G. W. G.
20. J. B. S. Haldane
21. O. F. Hall
22. Carl O. Hartman
23. Fowler V. Harper
24. Francis J. Hector
25. Norman E. Himes
26. B. H.
27. Sophia Kleegman
28. Robert Laidlaw
29. Karl S. Lashley
30. J. B. Lippincott [Howard K. Bauernfeind]
31. R. L.
32. Clyde E. Martin
33. William G. Mather Jr.
34. L. M.
35. C. W. Metz
36. Emily Hartshorne Mudd
37. John H. Mueller
38. Kate Hevner Mueller
39. Henry G. Nester
40. Howard M. Parshley
41. Glenn V. Ramsey
42. Thurman B. Rice
43. Albert R. Shadle
44. A. F. Shull

45. Lewis M. Terman
46. Ralph Voris
47. W. O. W.
48. D. and I. W.
49. Herman B Wells

Current Correspondence
1. American Museum of Natural History
2. Galleys of *Sexual Behavior in the Human Female*, 3N(2), boxes 1 and 2

Carney Landis Papers

Geography and Map Library, Indiana University, Bloomington, Indiana
Alfred C. Kinsey Map Collection

Lilly Library, Indiana University, Bloomington, Indiana
Muller, Hermann Joseph. Muller mss., 1910–1967
Weatherwax, Paul. Weatherwax mss., 1915–1975

Primary Sources

Aberle, Sophie D., and George W. Corner. *Twenty-Five Years of Sex Research: History of the National Research Council Committee for Research in Problems of Sex, 1922–1947.* Philadelphia: W. B. Saunders, 1952.

Abortion in the United States: A Conference Sponsored by the Planned Parenthood Federation of America, Inc. at Arden House and the New York Academy of Medicine. Edited by Mary Steichen Calderone. [New York]: Hoeber-Harper, [1958].

Achilles, Paul Strong. *The Effectiveness of Certain Social Hygiene Literature.* New York: American Social Hygiene Association, 1923.

Allen, Edgar, ed. *Sex and Internal Secretions: A Survey of Recent Research.* Baltimore: Williams & Wilkins, 1939.

Anderson, Edgar. "Hybridization in American *Tradescantias*." *Annals of the Missouri Botanical Garden* 23 (September 1936): 512–25.

———. "The Problem of Species in the Northern Blue Flags, *Iris versicolor* L. and *Iris virginica* L." *Annals of the Missouri Botanical Garden* 15 (September 1928): 241–313.

———. "Supra-specific Variation in Nature and in Classification from the View-Point of Botany." *American Naturalist* 71 (May–June 1937): 223–35.

Beach, Frank A. "A Cross-Species Survey of Mammalian Sexual Behavior." In Hoch and Zubin, *Psychosexual Development in Health and Disease*, 52–78.

———. "Effects of Injury to the Cerebral Cortex upon the Display of Masculine and Feminine Behavior by Female Rats." *Journal of Comparative Physiology* 36 (1943): 169–99.

———. *Hormones and Behavior: A Survey of Interrelationships between Endocrine Secretions and Patterns of Overt Response.* New York: Paul B. Hoeber, 1948.

———. "Hormones and Mating Behavior in Vertebrates." In *Recent Progress in Hor-*

mone Research, edited by Gregory Pincus, vol. 1, 27–63. New York: Academic Press, 1947.

———, ed. *Human Sexuality in Four Perspectives*. Baltimore: Johns Hopkins University Press, 1977.

———. "Karl Spencer Lashley: June 7, 1890–August 7, 1958." *Biographical Memoirs* 35 (1961): 163–204.

———. "A Review of Physiological and Psychological Studies of Sexual Behavior in Mammals." *Physiological Reviews* 27 (June 1947): 240–307.

———. "Sexual Behavior in Animals and Man." In *The Harvey Lectures: Delivered under the Auspices of the Harvey Society of New York, 1947–1948*. Vol. 43. Springfield, IL: C. C. Thomas, 1950.

Benedict, Ruth. *Patterns of Culture*. New York: Penguin, 1934.

Bernard, William S. "Student Attitudes on Marriage and the Family." *American Sociological Review* 3 (June 1938): 354–61.

Bigelow, Maurice A. *Sex-Education*. New York: Macmillan, 1916.

Black, Bertram J., and Edward B. Olds. "A Punched Card Method for Presenting, Analyzing, and Comparing Many Series of Statistics." *Journal of the American Statistical Association* 41 (1946): 347–55.

Blake, Robert Rogers, and Glenn V. Ramsey, eds. *Perception: An Approach to Personality*. New York: Ronald Press, 1951.

Blanchard, Phyllis, and Carlyn Manasses. *New Girls for Old*. New York: Macaulay, 1937.

Boas, Ernst P., and Ernst F. Goldschmidt. *The Heart Rate*. Springfield, IL: C. C. Thomas, 1932.

Bowman, Henry. "The Marriage Course at Stephens College." *Marriage and Family Living* 3 (February 1941): 8–9, 11.

———. "Report of Committee on College Courses in Preparation for Marriage." *Marriage and Family Living* 3 (May 1941): 36–37.

Bowman, Henry, Flora Thurston, and Margaret Wylie. "The Teacher as Marriage and Family Counselor [*sic*]." *Marriage and Family Living* 6 (November 1944): 76–78.

Bridges, Calvin B., and Thomas Hunt Morgan. "The Second-Chromosome Group of Mutant Characters." In *Contributions to the Genetics of Drosophila melanogaster*, 278 (Washington, DC: Carnegie Institute of Washington, 1919): 125–342.

Briffault, Robert. *The Mothers: A Study of the Origins of Sentiments and Institutions*. New York: Macmillan, 1931.

Brooks, JoAnn, and Helen C. Hofer, comps. *Sexual Nomenclature: A Thesaurus*. Boston: G. K. Hall & Co., 1976.

Bromley, Dorothy Dunbar, and Florence Haxton Britten. *Youth and Sex: A Study of 1300 College Students*. New York: Harper & Bros., 1938.

Burgess, Ernest W. "Three Pioneers in the Study of Sex." *Living* 1 (November 1939): 76, 95.

Burgess, Ernest W., and Leonard S. Cottrell Jr. "The Prediction of Adjustment in Marriage." *American Sociological Review* 1 (October 1936): 737–51.

Burgess, J. Stewart. "The College and the Preparation for Marriage and Family Relationships." *Living* 1 (May–August 1939): 39–42.

"The Bussey Institution of Harvard University: Founded 1872–Closed June 30, 1936." *Genetics* 21 (July 1936): 295–96.

Chapin, F. Stuart. *The Measurement of Social Status by the Use of the Social Status Scale 1933*. Minneapolis: University of Minnesota Press, 1933.

Clark, Leland C., and Paul Treichler. "Psychic Stimulation of Prostate Secretion." *Psychosomatic Medicine* 12 (July–August 1950): 261–63.

Cochran, William G., Frederick Mosteller, and John W. Tukey, *Statistical Problems of the Kinsey Report on* Sexual Behavior in the Human Male. Washington, DC: American Statistical Association, 1954.

———. "Statistical Problems of the Kinsey Report." *Journal of the American Statistical Association* 48 (December 1953): 676–713.

Darwin, Charles. *On the Origin of Species by Means of Natural Selection*. London: John Murray, 1859.

———. *The Descent of Man, and Selection in Relation to Sex*. London: John Murray, 1871.

Davis, Katharine Bement. *Factors in the Sex Life of Twenty-Two Hundred Women*. New York: Harper & Bros., 1929.

———. "Periodicity of Sex Desire." *American Journal of Obstetrics and Gynecology* 12 (December 1926): 824–38.

———. "Three Score Years and Ten: An Autobiographical Biography." *University of Chicago Magazine* 26 (December 1933): 58–61.

Deutsch, Helene. *The Psychology of Women: A Psychoanalytic Interpretation*. 2 vols. New York: Grune & Stratton, 1944–1945.

Dickinson, Robert Latou. *Human Sex Anatomy: A Topographical Hand Atlas*. Baltimore: Williams & Wilkins, 1933.

———. *Human Sex Anatomy: A Topographical Hand Atlas*. 2nd ed. Baltimore: Williams & Wilkins, 1949.

Dickinson, Robert Latou, and Lura Beam. *A Thousand Marriages: A Medical Study of Sex Adjustment*. Baltimore: Williams & Wilkins, 1931.

———. *The Single Woman: A Medical Study in Sex Education*. Baltimore: Williams & Wilkins, 1934.

Dobzhansky, Theodosius. "Biological Adaptation." *Scientific Monthly* 55 (November 1942): 391–402.

———. "A Critique of the Species Concept in Biology." *Philosophy of Science* 2 (July 1935): 344–55.

———. *Genetics and the Origin of Species.* New York: Columbia University Press, 1937.

———. "The Raw Materials of Evolution." *Scientific Monthly* 46 (May 1938): 445–49.

Dunn, Emmett Reid. *The Salamanders of the Family Plethodontidae.* Northampton, MA: Smith College, 1926.

Eckert, Wallace John. *Punched Card Methods in Scientific Computation.* New York: Thomas J. Watson Astronomical Computing Bureau, Columbia University, 1940.

Ellis, Havelock. *Psychology of Sex: A Manual for Students.* New York: Ray Long & Richard R. Smith, 1933.

———. *Studies in the Psychology of Sex.* 6 vols. Philadelphia: F. A. Davis, 1905–1913.

England, L. R. "Little Kinsey: An Outline of Sex Attitudes in Britain." *Public Opinion Quarterly* 13 (Winter 1949–50): 587–600.

Exner, Max Joseph. *Problems and Principles of Sex Education: A Study of 948 College Men.* New York: Association Press, 1915.

———. *The Sexual Side of Marriage.* New York: W. W. Norton, 1932.

Ferris, G. F. "Entomological Illustrations." *Science,* n.s., 58, no. 1501 (October 5, 1923): 265–66.

———. "The Place of the Systematist in Modern Biology." *Scientific Monthly* 16 (May 1923): 514–20.

Finger, Frank W. "Sex Beliefs and Practices among Male College Students." *Journal of Abnormal and Social Psychology* 42 (January 1947): 57–67.

"First Annual Meeting of the National Conference on Family Relations." *Living* 1 (January 1939): 30–31.

Fishbein, Morris, and Ernest W. Burgess, eds. *Successful Marriage.* Garden City, NY: Doubleday, 1947.

Flexner, Simon. "Jacques Loeb and His Period." *Science,* n.s., 66, no. 1711 (October 14, 1927): 333–37.

Folsom, Joseph K. "Changing Values in Sex and Family Relations." *American Sociological Review* 2 (October 1937): 717–26.

———, ed. *Plan for Marriage: An Intelligent Approach to Marriage and Parenthood; Proposed by Members of the Staff of Vassar College.* New York: Harper & Bros., 1938.

Ford, Clellan S. *A Comparative Study of Human Reproduction.* New Haven, CT: Yale University Press, 1945.

Ford, Clellan S., and Frank A. Beach. *Patterns of Sexual Behavior.* New York: Harper Colophon, 1951.

Freud, Sigmund. *A General Introduction to Psychoanalysis.* Translated by Joan Riviere. New York: Perma Giants, 1935.

———. *New Introductory Lectures on Psychoanalysis.* New York: W. W. Norton, 1933.

———. *Three Essays on the Theory of Sexuality.* New York: Basic Books, 1975.

"The Future Adventure." *Living* 1 (January 1939): 17–18.

Gantt, W. Horsley. *Experimental Basis for Neurotic Behavior.* New York: Paul B. Hoeber, 1944.

Gebhard, Paul H., and others of the Institute for Sex Research. *Pregnancy, Birth, and Abortion.* New York: Harper, 1958.

———. *Sex Offenders: An Analysis of Types.* New York: Harper & Row, 1965.

Gleason, H. A. "A Plea for Sanity in Nomenclature." *Science,* n.s., 71, no. 1844 (May 2, 1930): 458–59.

Goldschmidt, Richard. "*Cynips* and *Lymantria.*" *American Naturalist* 71 (September–October 1937): 508–14.

———. *The Material Basis of Evolution.* New Haven, CT: Yale University Press, 1940.

Groves, Ernest R. *Marriage.* New York: Henry Holt, 1933.

———. *The Marriage Crisis.* New York: Longmans, Green, 1928.

Groves, Ernest R., and Willam F. Ogburn. *American Marriage and Family Relationships.* New York: Henry Holt, 1928.

Hamilton, Gilbert V. *A Research in Marriage.* New York: Albert and Charles Boni, 1929.

———. "A Study of Sexual Tendencies in Monkeys and Baboons." *Journal of Animal Behavior* 4 (1914): 295–318.

Hamilton, Gilbert V., and Kenneth MacGowan. *What's Wrong with Marriage.* New York: Albert and Charles Boni, 1929.

Harvey, O. L. "The Questionnaire as Used in Recent Studies of Human Sexual Behavior." *Journal of Abnormal Psychology* 26, no. 4 (1932): 379–89.

Henry, Jules. "The Social Function of Child Sexuality in Pilaga Indian Culture." In Hoch and Zubin, *Psychosexual Development in Health and Disease,* 91–101.

Hitschmann, Eduard E., and Edmund Bergler. *Frigidity in Women: Its Characteristics and Treatment.* Translated by Polly Leeds Weil. Washington, DC: Nervous and Mental Disease Publishing Co., 1936.

Hoch, Paul H., and Joseph Zubin, eds. *Psychosexual Development in Health and Disease.* New York: Grune & Stratton, 1949.

Hohman, Leslie B., and Bertram Schaffner. "The Sex Life of Unmarried Men." *American Journal of Sociology* 52 (May 1947): 501–7.

Howson, Howard. "Emotional Maturity and the Approach to Marriage." In Folsom, *Plan for Marriage,* 58–71.

Hughes, Walter L. "Sex Experiences of Boyhood." *Journal of Mental Hygiene* 12 (May 1926): 262–73.

Huxley, Julian. *Evolution: The Modern Synthesis.* London: Harper & Bros., 1943.

"Indiana University Jordan Hall Dedication Program, June 6–9, 1956." http://www.bio.indiana.edu/about/history/JH_dedication_1956.pdf. Accessed July 20, 2013.

Jacob, Elin K. "Classification and Crossdisciplinary Communication: Breaching the Boundaries Imposed by Classificatory Structure." In *Advances in Knowledge Organization,* vol. 4, *Knowledge Organization and Quality Management,* edited by Hanne Albrechtsen and Susanne Oernager, 101–8. Frankfurt-am-Main: Indeks Verlag, 1994.

Jennings, Herbert Spencer. "Biographical Memoir of Raymond Pearl 1879–1940." *National Academy of Sciences Biographical Memoirs* 22 (1942): 295–347.

Johnson, Mary A. "A Course in Human Relations at Brooklyn College." *Living* 1 (November 1939): 73–74.

Jung, Moses. "The Course in Modern Marriage at the State University of Iowa." *Living* 1 (May–August 1939): 43, 50.

Katz, Leo. "Punched Card Technique for the Analysis of Multiple Level Sociometric Data." *Sociometry* 13, no. 2 (1950): 108–22.

Kempf, E. J. "The Social and Sexual Behavior of Infra-human Primates with Some Comparable Facts in Human Behavior." *Psychoanalytic Review* 4 (1917): 127–54.

Krafft-Ebing, Richard von. *Psychopathia Sexualis: With Especial Reference to Contrary Sexual Instinct; A Medico-Legal Study.* Philadelphia: F. A. Davis, 1892.

Kroger, William S., and S. Charles Freed. "Psychosomatic Aspects of Frigidity." *Journal of the American Medical Association* 143 (June 1950): 526–32.

Landis, Carney, and M. Marjorie Bolles. *Personality and Sexuality of the Physically Handicapped Woman.* New York: Paul B. Hoeber, 1942.

Landis, Carney, Agnes T. Landis, M. Marjorie Bolles, et al. *Sex in Development: A Study of the Growth and Development of the Emotional and Sexual Aspects of Personality together with Physiological, Anatomical, and Medical Information on a Group of 153 Normal Women and 142 Female Psychiatric Patients.* New York: Paul B. Hoeber, 1940.

Lashley, Karl S. "Mass Action in Cerebral Function." *Science,* n.s., 73, no. 1888 (March 6, 1931): 245–54.

———. "Persistent Problems in the Evolution of Mind." *Quarterly Review of Biology* 24 (March 1949): 28–42.

Lorge, Irving. "The 'Last School Grade Completed' as an Index of Intellectual Level." *School & Society* 56 (November 28, 1942): 529–32.

Lundberg, George A., and Pearl Friedman. "A Comparison of Some Measures of Socioeconomic Status." *Rural Sociology* 8 (1943): 227–42.

Lynd, Robert Staughton, and Helen Merrell Lynd. *Middletown: A Study in Contemporary American Culture.* New York: Harcourt, Brace, 1929.

———. *Middletown in Transition: A Study in Cultural Conflicts.* New York: Harcourt, Brace, 1937.

Malinowski, Bronislaw. *The Sexual Life of Savages in North-Western Melanesia: An Ethnographic Account of Courtship, Marriage, and Family Life among the Natives of the Trobriand Islands, British New Guinea.* New York: Halcyon House, 1929.

Masters, William H., and Virginia E. Johnson. *Human Sexual Response.* Boston: Little, Brown, 1966.

Matsner, Eric M. *The Technique of Contraception.* Baltimore: William Wood, 1936.

McNemar, Quinn. "Opinion-Attitude Methodology." *Psychological Bulletin* 43 (July 1946): 289–373.

Meagher, John F. W., and Smith Ely Jelliffe. *A Study of Masturbation and the Psychosexual Life.* 3rd ed. Baltimore: William Wood, 1936.

Merrill, Lilburn. "A Summary of Findings in a Study of Sexualism among a Group of One Hundred Delinquent Boys." *Journal of Delinquency* 3 (November 1918): 255–67.

Metcalf, C. L. *The Genitalia of Male Syrphidae.* Contributions from the Department of Zoology and Entomology 67. Columbus: Ohio State University, 1921.

Mickel, Clarence Eugene. "Biological and Taxonomic Investigations on the Mutillid Wasps." [*Smithsonian Institution*] *United States National Museum Bulletin* 143 (1928): 1–351.

Miller, Gerrit S., Jr. "The Primate Basis of Human Sexual Behavior." *Quarterly Review of Biology* 6 (December 1931): 379–410.

Money, John. "Hermaphroditism, Gender and Precocity in Hyperadrenocorticism: Psychologic Findings." *Bulletin of the Johns Hopkins Hospital* 96 (1955): 253–64.

Morgan, Thomas Hunt. *The Scientific Basis of Evolution.* New York: W. W. Norton, 1932.

———. "The Theory of the Gene." *American Naturalist* 51 (September 1917): 513–44.

Mowrer, Harriet R. "A Plan for Successful Marriages." *Living* 1 (May–August 1939): 47.

Muller, Hermann J. "Social Biology and Population Improvement." *Nature* 144 (September 16, 1939): 521.

Newcomb, Theodore. "Recent Changes in Attitudes toward Sex and Marriage." *American Sociological Review* 2 (October 1937): 659–67.

"Opinion: Sex or Snake Oil?" *Time,* January 11, 1954.

Osborn, Henry Fairfield. "Aristogenesis, the Creative Principle in the Origin of Species." *Science,* n.s., 79, no. 2038 (January 19, 1934): 41–45.

———. *The Origin and Evolution of Life: On the Theory of Action, Reaction, and Interaction of Energy.* 1917; New York: C. Scribner, 1925.

Parshley, Howard M. "Gall Wasps and the Species Problem." *Entomological News* 41 (June 1930): 191–95.

———. *The Science of Human Reproduction: Biological Aspects of Sex.* New York: Eugenics Publishing Co., 1933.

Pearl, Raymond. *The Biology of Population Growth.* New York: Alfred A. Knopf, 1925.

———. *Introduction to Medical Biometry and Statistics.* 3rd ed. Philadelphia: W. B. Saunders, 1940.

———. *Man the Animal*. Edited by Maud M. De Witt Pearl. Bloomington, IN: Principia Press, 1946.

———. *The Natural History of Population*. New York: Oxford University Press, 1939.

Pearl, Raymond, and W. Edwin Moffett. "Bodily Constitution and Human Longevity." *Proceedings of the National Academy of Sciences of the United States of America* 25 (December 15, 1939): 609–16.

Peck, Martin W., and F. Lyman Wells. "On the Psycho-sexuality of Graduate Men." *Mental Hygiene* 7 (October 1923): 697–714.

Peterson, Kenneth Martin. "Early Sex Information and Its Influence on Later Sex Concepts." MA thesis, University of Colorado, Boulder, 1938.

Popenoe, Paul. *Problems of Human Reproduction*. Baltimore: Williams & Wilkins, 1926.

———. "Mate Selection." *American Sociological Review* 2 (October 1937): 735–43.

———. *Preparing for Marriage*. Los Angeles: American Institute of Family Relations, 1938.

Ramsey, Glenn V. "Factors in the Sex Life of 291 Boys." EdD diss., Indiana University, Bloomington, 1941.

———. "The Sex Information of Younger Boys." *American Journal of Orthopsychiatry* 13 (April 1943): 347–52.

———. "The Sexual Development of Boys." *American Journal of Psychology* 56 (April 1943): 217–34.

Rice, Thurman B. *The Age of Romance*. Chicago: American Medical Association, 1933.

———. *How Life Goes On and On: A Story for Girls of High School Age*. Chicago: American Medical Association, 1933.

———. *The Human Body: Some Rules for Right Living*. New York: Funk & Wagnalls, 1937.

Sewell, William H. "The Development of a Sociometric Scale." *Sociometry* 5 (August 1942): 279–97.

Shadle, Albert R. "Copulation in the Porcupine." *Journal of Wildlife Management* 10 (April 1946): 159–62.

Shadle, Albert R., Marilyn S. Smelzer, and Margery M. Metz, "The Sex Reactions of Porcupines (*Erethizon d. dorsatum*) before and after Copulation." *Journal of Mammalogy* 27 (May 1946): 116–21.

Simpson, George Gaylord. "Supra-specific Variation in Nature and in Classification from the View-Point of Paleontology." *American Naturalist* 71 (May–June 1937): 236–67.

———. *Tempo and Mode in Evolution*. New York: Columbia University Press, 1944.

Snedecor, George W. *Statistical Methods Applied to Experiments in Agriculture and Biology*. 4th ed. Ames, IA: Collegiate Press, 1946.

Sorokin, Pitirim. *Social Mobility*. New York: Harper & Bros., 1927.

Squier, Raymond. "The Medical Basis of Intelligent Sexual Practice." In Folsom, *Plan for Marriage*, 113–37.

Stone, Calvin P. "Experimental Studies of Two Important Factors Underlying Masculine Sexual Behavior: The Nervous System and the Internal Secretion of the Testis." *Journal of Experimental Psychology* 6 (April 1923): 85–106.

———. "Further Study of Sensory Functions in the Activation of Sexual Behavior in the Young Male Albino Rat." *Journal of Comparative Psychology* 3 (December 1923): 469–73.

Stone, Hannah M., and Abraham Stone. *A Marriage Manual: A Practical Guide-Book to Sex and Marriage*. New York: Simon & Schuster, 1937.

———. *A Marriage Manual: A Practical Guide-Book to Sex and Marriage*. New York: Simon & Schuster, 1952.

Stopes, Marie. *Married Love*. 1918. Reprint, with a foreword by Ross McKibbin, Oxford: Oxford University Press, 2004.

Strakosh, Frances M. *Factors in the Sex Life of Seven Hundred Psychopathic Women*. Utica, NY: State Hospitals Press, 1934.

Taylor, William S. "A Critique of Sublimation in Males: A Study of Forty Superior Single Men." *Genetic Psychology Monographs* 13 (January 1933): 7–109.

———. *A Critique of Sublimation in Males*. Worcester, MA: Clark University, 1933.

Terman, Lewis M., "Kinsey's 'Sexual Behavior in the Human Male': Some Comments and Criticisms." *Psychological Bulletin* 45, no. 5 (1948): 443–59.

Terman, Lewis M., Paul Buttenwieser, Leonard W. Ferguson, Winifred Bent Johnson, and Donald P. Wilson. *Psychological Factors in Marital Happiness*. New York: McGraw-Hill, 1938.

Terman, Lewis M., and Winifred B. Johnson. "Methodology and Results of Recent Studies in Marital Adjustment." *American Sociological Review* 4 (June 1939): 307–24.

Terman, Lewis M., and Catharine C. Miles. *Sex and Personality: Studies in Masculinity and Femininity*. New York: McGraw-Hill, 1936.

Waller, Willard. "The Rating and Dating Complex." *American Sociological Review* 2 (October 1937): 727–34.

Walley, G. S. Review of *The Origin of Higher Categories in Cynips*, by Alfred C. Kinsey. *Canadian Entomologist* 69 (January 1937): 20–21.

Ward, Lester Frank. *Pure Sociology*. New York: Macmillan, 1903.

Warner, W. Lloyd, and Paul S. Lunt. *The Social Life of a Modern Community*. New Haven, CT: Yale University Press, 1941.

Wheeler, William Morton. "The Bussey Institution, 1871–1929." In *The Development of Harvard University since the Inauguration of President Eliot, 1869–1929*, edited by Samuel Eliot Morison, 508–17. Cambridge, MA: Harvard University Press, 1930.

———. "The Dry-Rot of Our Academic Biology." In *Foibles of Insects and Men*, 187–204. New York: Alfred A. Knopf, 1928.

———. "Insect Parasitism and Its Peculiarities." In *Foibles of Insects and Men*, 49–67. New York: Alfred A. Knopf, 1928.

———. "The Kelep Ant and the Courtship of Its Mimic, *Cardiacephala Myrmex*." In *Foibles of Insects and Men*, 147–65. New York: Alfred A. Knopf, 1928.

———. "The Organization of Research." *Science*, n.s., 53, no. 1360 (January 21, 1921): 53–67.

———. "The Physiognomy of Insects." In *Foibles of Insects and Men*, 3–45. New York: Alfred A. Knopf, 1928.

———. "The Termitodoxa, or Biology and Society." In *Foibles of Insects and Men*, 205–17. New York: Alfred A. Knopf, 1928.

Williams, Robert Hardin, ed. *Textbook of Endocrinology*. Philadelphia: W. B. Saunders, 1950.

Wilson, Pauline Park. "A Plan for Successful Marriage." *Living* 1 (January 1939): 8.

Wortis, S. Bernard. "The Premarital Interview." *Living* 2 (May 1940): 37–39.

Wright, Sewall. "The Role of Mutation, Inbreeding, Crossbreeding, and Selection in Evolution." *Proceedings of the Sixth International Congress of Genetics* 1 (1932): 356–66.

Wyman, Donald. "Notes of Interest." *Arnoldia* 3 (September 17, 1943): 44.

Zuckerman, Solly. *The Social Life of Monkeys and Apes*. London: Keegan Paul, Trench, Trubner, 1932.

Secondary Sources

Absher, Ruby Roten. "The History of the Indiana State Board of Health from 1922 to 1954." HSD diss., Indiana University, 1978.

Albright, Thomas G., ed. *The Kinsey Interview Kit Code Book*. Bloomington, IN: Kinsey Institute for Research in Sex, Gender, and Reproduction, 1991.

———, comp. and ed. *The Kinsey Interview Kit: Codebook*. 2nd ed. Bloomington, IN: Kinsey Institute for Research in Sex, Gender, and Reproduction, 2006.

Allen, Garland E. *Life Science in the Twentieth Century*. Cambridge: Cambridge University Press, 1979.

Allen, Judith A. "Kinsey's Women: Sexed Bodies, Heterosexualities, and Abortion, 1920–1960." Typescript, Department of Gender Studies, Indiana University, 2002.

———. "'The Overthrow' of Gynaecocentric Culture: Charlotte Perkins Gilman and Lester Frank Ward." In *Studies in American Literary Realism and Naturalism (Salrn)*, edited by Cynthia J. Davis and Denise D. Knight, 59–86. Tuscaloosa: University of Alabama Press, 2004.

———. "Testifying to Cold War Sexualities: Alfred Kinsey as Expert Witness, 1949–1955." Paper presented at the 13th Berkshire Conference on the History of Women, June 2002.

Allyn, David. "Private Acts/Public Policy: Alfred Kinsey, the American Law Institute, and the Privatization of American Sexual Morality." *Journal of American Studies* 30 (December 1996): 405–28.

Anscome, F. R. "Quiet Contributor: The Civic Career and Times of John W. Tukey." *Statistical Science* 18 (August 2003): 287–310.

Beatty, John. "Dobzhansky and the Biology of Democracy: The Moral and Political Significance of Genetic Variation." In *The Evolution of Theodosius Dobzhansky: Essays on His Life and Thought in Russia and America*, edited by Mark B. Adams, 195–218. Princeton, NJ: Princeton University Press, 1994.

Bailey, Beth. *From Front Porch to Backseat: Courtship in Twentieth Century America*. Baltimore: Johns Hopkins University Press, 1988.

———. "Scientific Truth . . . and Love: The Marriage Education Movement in the United States." *Journal of Social History* 20 (Summer 1987): 711–32.

———. *Sex in the Heartland*. Cambridge, MA: Harvard University Press, 1999.

Bancroft, John. "Alfred Kinsey's Work 50 Years Later." New introduction to *Sexual Behavior in the Human Female*, by Alfred C. Kinsey, Wardell B. Pomeroy, Clyde E. Martin, and Paul H. Gebhard, a–r. Bloomington: Indiana University Press, 1998.

Bashford, Alison, and Carolyn Strange. "Public Pedagogy: Sex Education and Mass Communication in the Mid–Twentieth Century." *Journal of the History of Sexuality* 13 (January 2004): 71–99.

Blair, Ann. *Too Much to Know: Managing Scholarly Information before the Modern Age*. New Haven, CT: Yale University Press, 2010.

Boyle, T. C. *The Inner Circle*. New York: Viking, 2005.

Brandt, Allan M. *No Magic Bullet: A Social History of Venereal Disease in the United States since 1880*. New York: Oxford University Press, 1985.

Brewer, Joan Scherer, ed. *The Kinsey Interview Kit*. Bloomington, IN: Kinsey Institute for Research in Sex, Gender, and Reproduction, 1985.

Brinkman, Paul Delbert. "Dr. Alfred C. Kinsey and the Press: Historical Case Study of the Relationship of the Mass Media and a Pioneering Behavioral Scientist." PhD diss., Indiana University, 1971.

Buhle, Mari Jo. *Feminism and Its Discontents: A Century of Struggle with Psychoanalysis*. Cambridge, MA: Harvard University Press, 1998.

Bullough, Vern. "Alfred Kinsey and the Kinsey Report: Historical Overview and Lasting Contributions." *Journal of Sex Research* 35 (May 1998): 127–31.

———. *Science in the Bedroom: A History of Sex Research*. New York: Basic Books, 1994.

Burian, Richard M. "How the Choice of Experimental Organism Matters: Epistemological Reflections on an Aspect of Biological Practice." *Journal of the History of Biology* 26 (Summer 1993): 351–67.

Burke, Peter. *A Social History of Knowledge: From Gutenberg to Diderot*. Cambridge, MA: Polity Press, 2000.

Butler, Judith. *Bodies That Matter: On the Discursive Limits of "Sex."* New York: Routledge, 1993.

———. "Imitation and Gender Insubordination." In *Inside/Out: Lesbian Theories, Gay Theories*, edited by Diana Fuss, 13–31. New York: Routledge, 1991.

Cain, Joe, ed. "Exploring the Borderlands: Documents of the Committee on Common Problems of Genetics, Paleontology, and Systematics, 1943–1944." *Transactions of the American Philosophical Society* 94, pt. 2 (2004): 1–160.

Cain, Joe, and Michael Ruse, eds. *Descended from Darwin: Insights into the History of Evolutionary Studies, 1900–1970*. Transactions of the American Philosophical Society, vol. 99, pt. 1. Philadelphia: American Philosophical Society, 2009.

Cain, Joseph Allen. "Common Problems and Cooperative Solutions: Organizational Activity in Evolutionary Studies, 1936–1947." *Isis* 84 (March 1993): 1–25.

———. "Rethinking the Synthesis Period in Evolutionary Studies." *Journal of the History of Biology* 42 (2009): 621–48.

Capshew, James H. *Herman B Wells: The Promise of the American University*. Bloomington: Indiana University Press, 2012.

Capshew, James H., Matthew H. Adamson, Patricia A. Buchanan, Narisara Murray, and Naoko Wake. "Kinsey's Biographers: A Historiographical Reconnaissance." *Journal of the History of Sexuality* 12 (July 2003): 465–86.

Carter, Julian B. "Birds, Bees, and Venereal Disease: Toward an Intellectual History of Sex Education." *Journal of the History of Sexuality* 10 (April 2001): 213–49.

———. *The Heart of Whiteness: Normal Sexuality and Race in America, 1880–1940*. Durham, NC: Duke University Press, 2007.

Chauncey, George, Jr. "From Sexual Inversion to Homosexuality: Medicine and the Changing Conceptualization of Female Deviance." *Salmagundi* 58–59 (Fall 1982–Winter 1983): 114–46.

Christenson, Cornelia V. *Kinsey, A Biography*. Bloomington: Indiana University Press, 1971.

Clark, Constance Areson. "Evolution for John Doe: Pictures, the Public, and the Scopes Trial Debate." *Journal of American History* 87 (March 2001): 1275–1303.

Clark, Thomas D. *Indiana University: Midwestern Pioneer*. Vol. 3, *Years of Fulfillment*. Bloomington: Indiana University Press, 1977.

Cooke, Kathy J. "From Science to Practice, or Practice to Science? Chickens and Eggs in Raymond Pearl's Agricultural Breeding Research, 1907–1916." *Isis* 88 (March 1997): 62–86.

———. "The Limits of Heredity: Nature and Nurture in American Eugenics before 1915." *Journal of the History of Biology* 31 (Summer 1998): 263–78.

Cott, Nancy F. *Public Vows: A History of Marriage and the Nation*. Cambridge, MA: Harvard University Press, 2000.

Cuordileone, K. A. *Manhood and American Political Culture in the Cold War*. New York: Routledge, 2004.

———. "'Politics in an Age of Anxiety': Cold War Political Culture and the Crisis in American Masculinity, 1949–1960." *Journal of American History* 87 (September 2000): 515–45.

Davis, Gregory K., Michael R. Dietrich, and David K. Jacobs. "Homeotic Mutants and the Assimilation of Developmental Genetics into the Evolutionary Synthesis, 1915–1952." In Cain and Ruse, *Descended from Darwin*, 133–54.

Davis, Rebecca L. *More Perfect Unions: The American Search for Marital Bliss*. Cambridge, MA: Harvard University Press, 2010.

———. "'Not Marriage at All, But Simple Harlotry': The Companionate Marriage Controversy." *Journal of American History* 94 (March 2008): 1137–63.

———. "'The Wife Your Husband Needs': Marriage Counseling, Religion, and Sexual Politics in the United States, 1930–1980." PhD diss., Yale University, 2006.

Del Tredici, Peter. "The Other Kinsey Report." *Natural History*, July–August 2006.

D'Emilio, John, and Estelle B. Freedman. *Intimate Matters: A History of Sexuality in America*. New York: Harper & Row, 1988.

———. "Problems Encountered in Writing the History of Sexuality: Sources, Theory, and Interpretation." *Journal of Sex Research* 27 (November 1990): 481–95.

Drucker, Donna J. "Beginning the Kinsey Reports: Glenn Ramsey's Sex Research in Peoria, 1938–1941." *Journal of Illinois History* 10 (Winter 2007): 271–88.

———. "'Building for a life-time of research': Letters of Alfred Kinsey and Ralph Voris." *Indiana Magazine of History* 106 (March 2010): 70–101.

———. "Creating the Kinsey Reports: Intellectual and Methodological Influences on Alfred Kinsey's Sex Research, 1919–1953." PhD diss., Indiana University, 2008.

———. "Keying Desire: Alfred Kinsey's Use of Punched-Card Machines for Sex Research." *Journal of the History of Sexuality* 22 (January 2013): 105–25.

———. *The Machines of Sex Research: Technology and the Politics of Identity, 1945–1985*. Dordrecht: Springer, 2014.

———. "Male Sexuality and Alfred Kinsey's 0–6 Scale: Toward 'A Sound Understanding of the Realities of Sex.'" *Journal of Homosexuality* 57, no. 9 (2010): 1105–23.

———. "Marking Sexuality from 0–6: The Kinsey Scale in Online Culture." *Sexuality and Culture* 16 (September 2012): 241–62.

———. "'A Most Interesting Chapter in the History of Science': Intellectual Responses to Alfred Kinsey's *Sexual Behavior in the Human Male*." *History of the Human Sciences* 25 (February 2012): 75–98.

———. "'A Noble Experiment': The Marriage Course at Indiana University, 1938–1940." *Indiana Magazine of History* 103 (September 2007): 231–64.

Edwards, Paul N., Lisa Gitelman, Gabrielle Hecht, Adrian Johns, Brian Larkin, and Neil Safier. "Historical Perspectives on the Circulation of Information." *American *Historical Review* 116 (December 2011): 1393–1435.

Engles, Eric William. "Biology Education in the Public High Schools of the United States from the Progressive Era to the Second World War: A Discursive History." PhD diss., University of California, Santa Cruz, 1991.

Ericksen, Julia A. with Sally A. Steffen. *Kiss and Tell: Surveying Sex in the Twentieth Century*. Cambridge, MA: Harvard University Press, 1999.

Evans, Mary Alice, and Howard Ensign Evans. *William Morton Wheeler, Biologist*. Cambridge, MA: Harvard University Press, 1970.

Fausto-Sterling, Anne. *Sexing the Body: Gender Politics and the Construction of Sexuality*. New York: Basic Books, 2000.

Feldstein, Ruth. *Motherhood in Black and White: Race and Sex in American Liberalism 1930–1965*. Ithaca, NY: Cornell University Press, 2000.

Fitzpatrick, Ellen. *Endless Crusade: Women Social Scientists and Progressive Reform*. New York: Oxford University Press, 1990.

Foucault, Michel. *The History of Sexuality: An Introduction*. Translated by Robert Hurley. Vol. 1. New York: Vintage, 1990.

———. *The Order of Things: An Archaeology of the Human Sciences*. New York: Vintage, 1970.

Galison, Peter, and Lorraine Daston. *Objectivity*. New York: Zone Books, 2007.

Garton, Stephen. *Histories of Sexuality: Antiquity to Sexual Revolution*. New York: Routledge, 2004.

Gathorne-Hardy, Jonathan. *Kinsey: Sex the Measure of All Things: A Life of Alfred C. Kinsey*. Bloomington: Indiana University Press, 2004.

Gathorne-Hardy, Jonathan, Linda Wolfe, and Bill Condon. *Kinsey: Public and Private*. New York: Newmarket Press, 2004.

Gebhard, Paul H., and Alan B. Johnson. *The Kinsey Data: Marginal Tabulations of the 1938–1963 Interviews Conducted by the Institute for Sex Research*. Philadelphia: W. B. Saunders, 1979.

Gerhard, Jane. *Desiring Revolution: Second Wave Feminism and the Rewriting of American Sexual Thought*. New York: Columbia University Press, 2001.

———. "Revisiting the 'Myth of the Vaginal Orgasm': The Female Orgasm in American Sexual Thought and Second Wave Feminism." *Feminist Studies* 26 (Summer 2000): 49–76.

Gilbert, James. *Men in the Middle: Searching for Masculinity in the 1950s*. Chicago: University of Chicago Press, 2005.

Gorchov, Lynn K. "Sexual Science and Sexual Politics: American Sex Research, 1920–1956." PhD diss., Johns Hopkins University, 2003.

Gould, Stephen Jay. "Of Wasps and WASPs." *Natural History* 91, no. 12 (December 1982): 8–15.

Griffith, R. Marie. "The Religious Encounters of Alfred Kinsey." *Journal of Ameri-*

can History 95 (September 2008): 349–77.
Hamilton, Andrew, and Quentin D. Wheeler. "Taxonomy and Why History of Science Matters for Science." *Isis* 99, no. 2 (2008): 331–40.
Hegarty, Peter. "Beyond Kinsey: The Committee for Research on Problems of Sex and American Psychology." *History of Psychology* 15, no. 3 (2012): 197–200.
———. *Gentlemen's Disagreement: Alfred Kinsey, Lewis Terman, and the Sexual Politics of Smart Men*. Chicago: University of Chicago Press, 2013.
———. "Getting Miles Away from Terman: Did the CRPS Fund Catharine Cox Miles's Unsilenced Psychology of Sex?" *History of Psychology* 15, no. 3 (2012): 201–8.
Heide, Lars. *Punched-Card Systems and the Early Information Explosion, 1880–1945*. Baltimore: Johns Hopkins University Press, 2009.
Igo, Sarah E. *The Averaged American: Surveys, Citizens, and the Making of a Mass Public*. Cambridge, MA: Harvard University Press, 2007.
Ilerbaig, Juan. "'The View-Point of a Naturalist': American Field Zoologists and the Evolutionary Synthesis, 1900–1945." In Cain and Ruse, *Descended from Darwin*, 23–48.
Irvine, Janice M. *Disorders of Desire: Sex and Gender in Modern American Sexology*. Philadelphia: Temple University Press, 1990.
———. "From Difference to Sameness: Gender Ideology in Sexual Science." *Journal of Sex Research* 27 (February 1990): 7–23.
Johnson, Colin R. *Just Queer Folks: Gender and Sexuality in Rural America*. Philadelphia: Temple University Press, 2013.
———. "Columbia's Orient: Gender, Geography, and the Invention of Sexuality in Rural America." PhD diss., University of Michigan, 2003.
Johnson, David K. *The Lavender Scare: The Cold War Persecutions of Gays and Lesbians in the Federal Government*. Chicago: University of Chicago Press, 2004.
Jones, James H. *Alfred C. Kinsey: A Public/Private Life*. New York: W. W. Norton, 1997.
———. "The Origins of the Institute for Sex Research: A History." PhD diss., Indiana University, 1973.
Kenen, Stephanie Hope. "Scientific Studies of Human Sexual Difference in Interwar America." PhD diss., University of California, Berkeley, 1998.
———. "Who Counts When You're Counting Homosexuals? Hormones and Homosexuality in Mid-Twentieth-Century America." In *Science and Homosexualities*, edited by Vernon A. Rosario, 197–218. New York: Routledge, 1997.
Klein, Julie Thompson. *Interdisciplinarity: History, Theory, and Practice*. Detroit: Wayne State University Press, 1990.
Kleinman, Kim. "Biosystematics and the Origin of Species: Edgar Anderson, W. H. Camp, and the Evolutionary Synthesis." In Cain and Ruse, *Descended from Darwin*, 73–91.

———. "His Own Synthesis: Corn, Edgar Anderson, and Evolutionary Theory in the 1940s." *Journal of the History of Biology* 32, no. 2 (Fall 1999): 293–320.

Knopf, Olga. *The Art of Being a Woman*. Translated by Alan Porter. Boston: Little, Brown, 1932.

Kohler, Robert E. *All Creatures: Naturalists, Collectors, and Biodiversity, 1850–1950*. Princeton: Princeton University Press, 2006.

———. "*Drosophila* and Evolutionary Genetics: The Moral Economy of Scientific Practice." *History of Science* 29 (1991): 335–75.

———. "*Drosophila:* A Life in the Laboratory." *Journal of the History of Biology* 26 (Summer 1993): 281–310.

———. *Landscapes and Labscapes: Exploring the Lab-Field Border in Biology*. Chicago: University of Chicago Press, 2002.

———. *Lords of the Fly:* Drosophila *Genetics and the Experimental Life*. Chicago: University of Chicago Press, 1994.

Krajewski, Markus. *Paper Machines: About Cards and Catalogs, 1548–1929*. Translated by Peter Knapp. Cambridge, MA: MIT Press, 2011.

Krementsov, Nikolai. "A Particular Synthesis: Aleksander Promptov and Speciation in Birds." *Journal of the History of Biology* 40 (2007): 637–82.

Krich, Aron. "Before Kinsey: Continuity in American Sex Research." *Psychoanalytic Review* 53 (Summer 1966): 69–90.

Kunzel, Regina. *Fallen Women, Problem Girls: Unmarried Mothers and the Professionalization of Social Work, 1890–1945*. New Haven, CT: Yale University Press, 1995.

Laporte, Léo F. "George G. Simpson, Paleontology, and the Expansion of Biology." In *The Expansion of American Biology*, edited by Keith R. Benson, Jane Maienschein, and Ronald Rainger, 80–107. New Brunswick: Rutgers University Press, 1991.

Latour, Bruno. *Science in Action: How to Follow Scientists and Engineers throughout Society*. Cambridge, MA: Harvard University Press, 1987.

Latour, Bruno, and Steve Woolgar. *Laboratory Life: The Construction of Scientific Facts*. Princeton: Princeton University Press, 1986.

Levens, Joshua P. "Sex, Neurosis, and Animal Behavior: The Emergence of American Psychobiology and the Research of W. Horsley Gantt and Frank A. Beach." PhD diss., Johns Hopkins University, 2005.

Lewis, Carolyn Herbst. *Prescription for Heterosexuality: Sexual Citizenship in the Cold War Era*. Chapel Hill: University of North Carolina Press, 2010.

———. "*Coitus Perfectus*: The Medicalization of Heterosexuality in the Cold War United States." PhD diss., University of California, Santa Barbara, 2007.

Light, Jennifer S. "When Computers Were Women." *Technology and Culture* 40, no. 3 (1999): 455–83.

Lloyd, Elisabeth A. *The Case of the Female Orgasm: Bias in the Science of Evolution*. Cambridge, MA: Harvard University Press, 2005.

Malcolm, Noel. "Thomas Harrison and His 'Ark of Studies': An Episode in the History of the Organization of Knowledge." *Seventeenth Century* 19 (Autumn 2004): 196–232.

May, Elaine Tyler. *Homeward Bound: American Families in the Cold War Era*. New York: Basic Books, 1988.

Mayr, Ernst. "The Role of Systematics in the Evolutionary Synthesis." In *The Evolutionary Synthesis: Perspectives on the Unification of Biology*, edited by Ernst Mayr and William B. Provine, 123–36. Cambridge, MA: Harvard University Press, 1980.

McLaren, Angus. *Twentieth-Century Sexuality: A History*. Oxford: Blackwell, 1999.

McWhirter, Donald P., Stephanie A. Sanders, and June M. Reinisch, eds. *Homosexuality/Heterosexuality: Concepts of Sexual Orientation*. New York: Oxford University Press, 1990.

Meyerowitz, Joanne. "Beyond the Feminine Mystique: A Reassessment of Post-war Mass Culture, 1946–1958." *Journal of American History* 79 (March 1993): 1455–82.

———. *How Sex Changed: A History of Transsexuality in the United States*. Cambridge, MA: Harvard University Press, 2002.

———, ed. *Not June Cleaver: Women and Gender in Postwar America, 1945–1960*. Philadelphia: Temple University Press, 1994.

———. "Sex Research at the Borders of Gender: Transvestites, Transsexuals, and Alfred C. Kinsey." *Bulletin of the History of Medicine* 75 (Spring 2001): 72–90.

Milam, Erika Lorraine. "'The Experimental Animal from the Naturalist's Point of View': Behavior and Evolution at the American Museum of Natural History, 1928–1954." In Cain and Ruse, *Descended from Darwin*, 157–78.

Minton, Henry L. *Departing from Deviance: A History of Homosexual Rights and Emancipatory Science in America*. Chicago: University of Chicago Press, 2002.

———. *Lewis M. Terman: Pioneer in Psychological Testing*. New York: New York University Press, 1988.

Morantz, Regina Markell. "Scientist as Sex Crusader: Alfred C. Kinsey and American Culture." *American Quarterly* 29 (Winter 1977): 145–66.

Neuhaus, Jessamyn. "The Importance of Being Orgasmic: Sexuality, Gender, and Marital Sex Manuals in the United States, 1920–1963." *Journal of the History of Sexuality* 9 (April 2000): 447–73.

Norberg, Arthur L. 1990. "High-Technology Calculation in the Early 20th Century: Punched Card Machinery in Business and Government." *Technology and Culture* 31, no. 4 (1990): 753–79.

Pagni, Charlotte F. "Hollywood Does Kinsey: Cinema, Sexology, and Cultural Regulation, 1948–1968." PhD diss., University of Michigan, 2003.

Passet, Joanne. *Sex Variant Woman: The Life of Jeannette Howard Foster*. Cambridge, MA: Da Capo Press, 2008.

Pauly, Philip J. *Biologists and the Promise of American Life: From Meriwether Lewis to Alfred Kinsey*. Princeton: Princeton University Press, 2000.

Pettit, Michael. "The Queer Life of a Lab Rat." *History of Psychology* 15, no. 3 (2012): 217–27.

Pivar, David J. *Purity and Hygiene: Women, Prostitution, and the "American Plan," 1900–1930*. Westport, CT: Greenwood, 2002.

Pomeroy, Wardell B. *Dr. Kinsey and the Institute for Sex Research*. New York: Harper & Row, 1972.

———. *Dr. Kinsey and the Institute for Sex Research*. 2nd ed. New Haven, CT: Yale University Press, 1982.

Provine, William B. *The Origins of Theoretical Population Genetics*. Chicago: University of Chicago Press, 1971.

Pryce, Anthony. "Let's Talk about *Sexual Behavior in the Human Male*: Kinsey and the Invention of (Post)Modern Sexualities." *Sexuality & Culture* 10 (Winter 2006): 63–93.

Rader, Karen A. "'The Mouse People': Murine Genetics Work at the Bussey Institution, 1909–1936." *Journal of the History of Biology* 31 (September 1998): 327–54.

Rainger, Ronald, Keith R. Benson, and Jane Maienschein, eds. *The American Development of Biology*. Philadelphia: University of Pennsylvania Press, 1988.

Richard, G. "The Historical Development of Nineteenth and Twentieth Century Studies on the Behavior of Insects." In *History of Entomology*, edited by Ray F. Smith, Thomas E. Mittler, and Carroll N. Smith, 477–502. Palo Alto, CA: Annual Reviews, 1973.

Roe, Anne. *The Making of a Scientist*. New York: Dodd, Mead, & Company, 1952.

Rosenberg, Daniel. "Early Modern Information Overload." *Journal of the History of Ideas* 64 (January 2003): 1–9.

Reumann, Miriam G. *American Sexual Character: Sex, Gender, and National Identity in the Kinsey Reports*. Berkeley: University of California Press, 2005.

———. "American Sexual Character in the Age of Kinsey, 1946–1964." PhD diss., Brown University, 1998.

Robinson, Paul. *The Modernization of Sex: Havelock Ellis, Alfred Kinsey, William Masters, and Virginia Johnson*. Ithaca, NY: Cornell University Press, 1989.

Rosenzweig, Louise Ritterskamp, and Saul Rosenzweig. "Notes on Alfred C. Kinsey's Pre-sexual Scientific Work and the Transition." *Journal of the History of the Behavioral Sciences* 5 (April 1969): 173–81.

Ross, Herbert H. "Evolution and Phylogeny." In *History of Entomology*, edited by Ray F. Smith, Thomas E. Mittler, and Carroll N. Smith, 171–84. Palo Alto, CA: Annual Reviews, 1973.

Sax, Karl. "The Bussey Institution." *Arnoldia* 7 (April 4, 1947): 13–16.

———. "The Bussey Institution: Harvard University Graduate School of Applied Biology, 1908–1936." *Journal of Heredity* 57 (1966): 175–79.

Scott, Joan W. "The Evidence of Experience." *Critical Inquiry* 17 (Summer 1991): 773–97.

———. "Gender: A Useful Category of Historical Analysis." *American Historical Review* 91 (December 1986): 1053–75.

Serlin, David. "Carney Landis and the Psychosexual Landscape of Touch in Mid-Twentieth-Century America." *History of Psychology* 15, no. 3 (2012): 209–16.

Shera, Jesse H. *Libraries and the Organization of Knowledge*. Hamden, CT: Archon Books, 1965.

Sleigh, Charlotte. *Six Legs Better: A Cultural History of Myrmecology*. Baltimore: Johns Hopkins University Press, 2007.

Smith, Barry, and Bert Klagges. "Philosophy and Biomedical Information Systems." In *Applied Ontology: An Introduction*, edited by Katherine Munn and Barry Smith, 21–38. Frankfurt: Ontos Verlag, 2008.

Smocovitis, Vassiliki Betty. "Organizing Evolution: Founding the Society for the Study of Evolution (1939–1950)." *Journal of the History of Biology* 27 (Summer 1994): 241–309.

———. *Unifying Biology: The Evolutionary Synthesis and Evolutionary Biology*. Princeton: Princeton University Press, 1996.

Spring, Justin. *Secret Historian: The Life and Times of Samuel Steward, Professor, Tattoo Artist, and Sexual Renegade*. New York: Farrar, Straus, and Giroux, 2010.

Stapleford, Thomas A. *The Cost of Living in America: A Political History*. Cambridge, MA: Cambridge University Press, 2009.

———. "Market Visions: Expenditure Surveys, Market Research, and Economic Planning in the New Deal." *Journal of American History* 94 (September 2007): 418–44.

Stern, Alexandra Minna. "'We Cannot Make a Silk Purse out of a Sow's Ear': Eugenics in the Hoosier Heartland." *Indiana Magazine of History* 103 (March 2007): 3–38.

Stevens, Kenneth R. "*United States vs. 31 Photographs*: Dr. Alfred C. Kinsey and Obscenity Law." *Indiana Magazine of History* 71 (December 1975): 299–318.

Stein, Marc. "The City of Sisterly and Brotherly Loves: The Making of Lesbian and Gay Movements in Greater Philadelphia, 1948–72." PhD diss., University of Pennsylvania, 1994.

Suresha, Ron Jackson. "'Properly Placed before the Public': Publication and Translation of the Kinsey Reports." *Journal of Bisexuality* 8, nos. 3–4 (2008): 203–28.

Terry, Jennifer. *An American Obsession: Science, Medicine, and Homosexuality in Modern Society*. Chicago: University of Chicago Press, 1999.

Turner, Christopher. "A Lot of Gall." *Cabinet* 25, Spring 2007.

Wallace, Irving. *The Chapman Report*. New York: Simon & Schuster, 1960.

Weir, J. A. "Harvard, Agriculture, and the Bussey Institution." *Genetics* 136 (1994): 1227–31.

Wells, Herman B. *Being Lucky: Reminiscences and Reflections.* Bloomington: Indiana University Press, 1980.
Yamashiro, Jennifer P. "Sex in the Field: Photography at the Kinsey Institute." PhD diss., Indiana University, 2002.
Yates, JoAnne. "Co-Evolution of Information-Processing Technology and Use: Interaction between the Life and Tabulating Industries." *Business History Review* 67 (1993): 1–51.
Yudell, Michael. "Kinsey's *Other* Report." *Natural History* 108 (July–August 1999): 80–81.
Zerubavel, Eviatar. *The Fine Line: Making Distinctions in Everyday Life.* Chicago: University of Chicago Press, 1993.
Zuk, Marlene. *Sexual Selections: What We Can and Can't Learn about Sex from Animals.* Berkeley: University of California Press, 2003.

Films

Alfred C. Kinsey [: A Documentary]. Toronto: CityTV, 2001.
The Chapman Report. Directed by George Cukor. Los Angeles: Darryl F. Zanuck Productions, 1962.
Kinsey. Directed by Bill Condon. Los Angeles: Fox Searchlight Films, 2004.
Kinsey. Directed by Barak Goodman and John Maggio. Boston: WGBH, PBS American Experience, 2005.
Reputations: Alfred Kinsey. Directed by Clare Beavan. London: BBC TV, 1996.
Sex and the Scientist. Directed by Jim Bales. Bloomington: Indiana University Audio-Visual Center, 1989.

Index

Note: Page numbers in italic refer to figures.

accumulative incidence curve, 113
adolescence, 134; same-sex relationships in, 136–37; sex play in, 152–53
age, relation to sexual behavior, 112–13, 133, 147
American Association for the Advancement of Science (AAAS), 60
American Museum of Natural History, 31, 33
American Social Hygiene Association, 142
American Statistical Association (ASA), team reviewing Kinsey's statistical methods, 12, 196n39; criticisms by, 103, 129, 145–47
anatomy: clitoral *vs.* vaginal orgasm, 156–57; male-female similarities and differences in, 148–49, 155–57
Anderson, Edgar, 18, 29, 35, 43, 51, 53; on Kinsey's gall wasp collection, 25, 27–28, 31; on Kinsey's sex research, 61, 87
animals, 153, 157; human sexual behavior *vs.*, 72, 144, 150–51; influences on sexual behavior of, 158–60; petting and foreplay in, 152, 156; same-sex relations among, 154–55; sexual behavior of, 155–56, 158
anonymity, of sex interview forms, 110
anthropological data, in *Female* volume, 150–53
arousal, 156–57, 160–61
art, Kinsey's classification system for, 2
"attic films," 155

Banks, Nathan, 33
Beach, Frank A., 153–54, 203n28; on animals' sexual behavior, 156–57; influence on Kinsey, 150, 161–62; on male-female differences in sexual behavior, 157–58, 160–61, 168
Beam, Lura, 48, 119

Bigelow, Maurice, 49–50
biographies, of Kinsey, 6
biology, 56, 102; changes in, 51, 56–61; choice of study species in, 35–36, 58–59; increasing focus on evolutionary synthesis in, 11, 51; Kinsey in histories of, 6–7; Kinsey's interest in education in, 15, 19–20, 42–50; Kinsey's training in, 15–16; Kinsey teaching, 61–63; Kinsey teaching future teachers of, 14, 64–65, 68–69
biometry, 102–3, 196n41; Kinsey's use of, 103, 122
Blair, Albert P. "Pat," 170
Bowdoin College, Kinsey at, 14, 15–16
Bowman, Henry, 78
brain, influence on sexual behavior, 160–62
Brown, Jean, *93*
Brues, Charles T., 16, 21–22, 25, 34
Bugbee, Robert, 26–27
Bureau of Social Hygiene, 90
Bussey Institution, *17*; Kinsey sending collected gall wasps back to, 24–25; Kinsey training at, 10–11, 15–16, 22–24; rough conditions at, 16, 18; taxonomy taught at, 16–17, 21–22

Cain, Joe: on biology, 51, 59; on Kinsey, 28, 56, 62
camps, Kinsey working at boys', 14
Carpenter, Frank, 23
causation, 22; Kinsey on, 9–10, 103, 134
celibacy, as abnormal, 135
Chapin, F. Stuart, 126
children: same-sex relationships of, 136–37; sex play among, 152–53; sexual behavior of, 95–96, 130, 167; social class and, 126, 128–29
Christenson, Carroll, 192n53
Christenson, Cornelia V., 6, 101
class, social, 198n76; clinicians redirecting behavior toward norms of, 130–31; data dropped from *Female* volume, 12, 146;

235

class, social (*cont.*); influence on sexual behavior, 122, 125, 127–32; Kinsey defining levels of, 126–27, 146, 200n30; stability and change in, 128–30

classification: data interpretation and, 8, 134; in *Edible Wild Plants of Eastern North America*, 40–42; in *Female* volume, 143–44; of gall wasps, 24–27; ill-effects of, 130, 132; importance of individuals in, 23–24; importance of looking at whole organism in, 23, 25; Kinsey challenging accepted forms of, 38, 166; Kinsey's skill in, 7–9, 166–67; in *Male* volume, 116–17, 143; not necessarily hierarchical, 42, 166; separation of parts in, 21–22; in sex research, 119, 144, 169, 170; shaping Kinsey's academic life, 7, 165–66; significant *vs.* insignificant differences in, 21–22; useful *vs.* non-useful, 40–42

Cochran, William I., 146, 196n39

coding: in Kinsey's labeling and storage of gall wasps, 29; on sex history interview forms, 29, 94, 109–10

collecting, 26; Kinsey's passion for, 1, 7–8, 29, 179n41. *See also* data collection

Committee for Research in Problems of Sex (CRPS), 90, 110

conditionability, Kinsey attributing male-female sexual differences to, 13, 41, 157–58, 160

consent, as an issue in sex history interviews, 190n32

contraception, 75–77, 162

"Criteria for a Hormonal Explanation of the Homosexual," 136–37

"Critique of Sublimation in Males" (Taylor), 70–71

culture, 162; gender inequality in, 153–54, 168; influence on sexual behavior, 83, 161–62; sexual myths in, 147; treatment of homosexuality in, 154

Cynips. See gall wasps

Daston, Lorraine, 26, 31

data analysis: distinguishing significant from insignificant differences, 21–22; enabled by punched-card machines, 132, 133, 137, 140; for information from sex history interviews, 98; Kinsey rejecting on gall wasps, 102; Kinsey's methods of, 22, 88, 107, 116; necessity of organization for, 165–66; of previous sex researchers, 107; sample size's relation to conclusions from, 118–19; in sex research, 49, 88, 120, 197n71. *See also* punched-card machines

data collection, 56, 119; on gall wasps, 24–29, 35, 51–52; Kinsey Reports' explanation of, 119–20, 145–46; Kinsey's, 13, 17, 179n41; relation to classification methods, 8; from sex history interviews, 12, 63–64, 78–79, 91; in sex research, 88, 114–15

data interpretation, 132; criticisms of *Male* volume's, 129–30, 133–34; in *Male* volume, 125, 128–29

data manipulation, 29; from coded sex history interview forms, 122–23; Kinsey's, 8, 9; Kinsey's horizontal scales, 9, 12; with punched-card machines, 110–14, 111

date of birth, influence on women's sexual behavior, 147

Davis, Katharine Bement, 48; influence on Kinsey, 70, 90, 136; on orgasms, 96, 120; as resource for marriage course, 70, 74; sex history surveys used by, 90–92

Dellenback, Bill, 155

Dickinson, Robert Latou, 48, 73–74, 90, 119

Dobzhansky, Theodosius, 65–66

Dunn, Emmett Reid, 35

Dunn, Leslie C., 22, 65–66

East, Edward M., 20, 22–23

Edible Wild Plants of Eastern North America (Kinsey and Fernald), 2, 11, 16, 38–42, 94

education, 22; influence on sexual behavior, 112–13, 129, 131, 133; Kinsey's interest in biology, 15, 42–50; Kinsey's pedagogical values, 11, 20, 43–44, 61, 64–67; Kinsey's teaching, 11, 14, 26–27, 61–62, 64, 66, 68; social class and, 126–27, 200n30. *See also* sex education

education, Kinsey's, 14–15, 33–35; training in entomology, 10–11; training in taxonomy, 8, 21–24
Ellis, Havelock, 6, 49–50
Emerson, Alfred E., 60, 116
entomology, 20; American Museum of Natural History's collection of, 31–33; Kinsey as leading entomologist on gall wasps, 25, 33, 51; Kinsey in, 4–5, 58, 65; Kinsey's professional relationships in, 15, 31; Kinsey's training in, 10–11, 14, 16, 34; methods in, 29, 34; taxonomy in, 21–24. *See also* gall wasps
Ericksen, Julia A., 6
eugenics, 158, 206n69; Kinsey not advocating, 19, 21
evolution, 185n48; criticisms of Kinsey's theories on, 5, 50–56; focusing on processes *vs.* individuals, 59–60; of gall wasps, 33, 35–36; gall wasps in Kinsey's "higher categories" concept, 54–55, 57; Kinsey on, 11, 38; in Kinsey's biology textbooks, 42–50; Kinsey teaching, 61, 64–67
Evolution (journal), 60
evolutionary synthesis, 2, 11, 51, 59
evolution studies, 58; changes in, 2, 11, 36–37, 56; Kinsey and, 51–52, 60; panel on supraspecific variation in, 53–56; speciation in, 36, 60
Exner, Max Joseph, 97–98
extramarital sex, in *Female* volume, 153–54
extrapolation, Kinsey's caution about, 103

Factors in the Sex Life of Twenty-Two Hundred Women (Davis), 70, 74, 90
fantasies, in men's and women's sex lives, 148
Fausto-Sterling, Anne, 9, 160–61
Female volume. See *Sexual Behavior in the Human Female*
feminism, Kinsey's role in, 6
Fernald, Merritt Lydon, 16, 25; Kinsey coauthoring book with, 2, 16, 39–42
Field and Laboratory Manual in Biology, 46
fieldwork: biologists turning to laboratory work over, 2, 36–37, 56–58; Kinsey's gall wasps collection trips, 24–27; Kinsey's preference for, 21, 34, 53; laboratory work *vs.*, 19–20, 58, 64; seen as masculine, 11, 20
filming, of sexual behavior, 155, 164
Folsom, Joseph Kirk, 78
Ford, Clellan S., 153–54, 157
Freud, Sigmund, 4, 49–50, 107, 156–57

Galison, Peter, 26, 31
galls, Kinsey's drawings of, *30*
"The Gall Wasp Genus *Cynips:* A Study in the Origin of Species," 35, 51
gall wasps, 22; criticisms of Kinsey's theories on, 50–56, 166–67; evolution of, 11, 33, 35–36, 50–56; Kinsey as leading entomologist on, 25, 51, 164; Kinsey identifying new species of, 2, 33, 51–52; Kinsey rejecting statistical analysis of, 102; Kinsey's classifications of, 2, 24, 166–67; Kinsey's collecting trips for, 24–27, 35; Kinsey's collection of, 27–28, 31, 33, 35, 51; Kinsey's dissertation on, 24, 33–35; Kinsey's drawings of, *32*; Kinsey's field notebooks on collection of, 28–29; in Kinsey's "higher categories" concept, 54–55, 57; Kinsey's move to sex research from, 2, 4–5, 11, 21; Kinsey's research on, 5, 16, 38, 124–25; limitations as study species, 24, 58; as study species, 35–36, 58
Gathorne-Hardy, Jonathan, 6
Gavit, Bernard, 192n53
Gebhard, Paul H., 33, 94, 110; on punched-card machines, 112, 197n71; on selection of data, 145, 197n71
gender, use of term, 167
gender bias, 10
gender differences. See male-female differences and similarities
genetics, 20; Bussey Institution training students in, 16–17; taxonomy and, 53, 60
Gerhard, Jane, 6
Goldschmidt, Richard, 20, 51, 56
Gorchov, Lynn K., 6
Gould, Stephen Jay, 56
Green, Kenneth S., 95
Gregory, William K., 53–55
Groves, Ernest, 78

Haldane, J. B. S., 61
Hall, Esther, 22
Hamilton, Gilbert V., 48, 72, 92, 96, 120
Harper, Fowler V., 192n53
Harvard University: Museum of Comparative Zoology at, 33. *See also* Bussey Institution
Hegarty, Peter, 131
heterosexuality. *See also* 0–6 scale
Hewitt, C. E., 16
historians, organization of data by, 5–6
historiography, of Kinsey, 6–7
Holland, J. E. P., 69
Hollerith machines. *See* punched-card machines
homosexuality: in animals, 151, 154–55; as centerpiece of Kinsey's thinking, 130, 169; in Davis's survey, 90–92; Kinsey's, 178n22; Kinsey's criticism of treatment in other texts, 49–50; Kinsey's interest in, 79–80; *Male* volume on, 127–32, 136–40; as normal, 85; prevalence of, 127–32, 140, 169. *See also* 0–6 scale
hormones: influence on sexual behavior, 148–49, 158–60; sexual identity and, 136–37
humans: as Kinsey's new study species, 58–59, 63; variations in behaviors of, 100–101
human sciences: biology applied to, 19, 68; in Kinsey's biology textbooks, 42–43; quantitative methods in, 114, 116, 133; scientists appropriate to research, 49, 130; taxonomy applied to, 117–19; Wheeler on biologists' perspective on, 11, 15, 18–20
Human Sex Anatomy (Dickinson), 73–74, 76
Huxley, Julian, 61

Igo, Sarah E., 6, 97, 169
Indiana University (IU): dedication of Jordan Hall at, 170, 207n11; Kinsey as professor at, 31, 61–67; Kinsey's marriage course at, 62, 67–86; Pearl's Patten lectures at, 98–100
individuals, 103; biology focusing on processes vs., 59–60; importance of, 23–24, 53, 100

insect behavior, Wheeler's interest in, 17
Institute for Sex Research (ISR), 2, 6, 110, 155
An Introduction to Biology, 1, 42–43
Introduction to Medical Biometry and Statistics (Pearl), 101, 106
Irvine, Janice M., 6

Jelliffe, Smith Ely, 49–50
Jenkins, J. A., 60
Johnson, Alan B., 197n71
Johnson, Colin R., 7
Johnson, Virginia E., 6, 114
Jones, James H., 6

Keisler, June Hiatt, 28
Kilfoil, Thomas, 192n53
Kinsey, Alfred, *27, 93, 113*; historiography of, 6–7; works of, 38 (*See also specific titles*)
Kinsey, Clara, 33, 164
Kinsey Reports, 21, 112, 114, 202n9; audiences for, 130–31; comparing animal and human sexual behavior, 99–100; criticisms of, 9–10; difference from previous sex research literature, 107; differences between *Male* and *Female* volumes, 12–13; influence of, 9, 169, 170; nine volumes planned for, 142–43, 147; selection of data for, 94, 146–47. *See also Sexual Behavior in the Human Female*; *Sexual Behavior in the Human Male*
Kleegman, Sophia, 156–57
Kohler, Robert E., 7–8, 22, 34, 56
Kohlmeier, Albert L., 192n53
Krafft-Ebing, Richard von, 4, 49–50, 107
Kroc, Robert L., 71, 76–77, 101

laboratory animals, increasing use of, 2
laboratory work. *See under* fieldwork
Landis, Carney, 96, 120
Lashley, Karl S., 147–48, 156, 161
Latour, Bruno, 7, 8, 112, 197n66
Leven, Joshua P., 6–7
libraries: importance of classification in, 3, 9; on relation of classification and data gathering methods, 8

Libraries and the Organization of Knowledge (Shera), 3
library, ISR, 2
life sciences, 15, 26; changes in, 4–5, 58–59; Kinsey and, 14, 15, 50, 58. *See also* biology
Lloyd, Elisabeth A., 199n15
Locke, Harvey, 77
love, and sex, 82–83
Lunt, Paul S., 4, 126, 128

Macfarlane, John Muirhead, 60
male-female differences and similarities, 135, 146; in arousal and orgasm, 157, 199n15; on extramarital sex, 153–54; *Female* volume focusing on, 143, 145, 147–48, 167; hierarchies and, 166; on homosexuality, 154; individual variations and, 162–63; Kinsey challenging gender dichotomy, 2, 144, 162–63; Kinsey evidence on, 149; Kinsey not seeing gender inequality, 168; Kinsey seeking reasons for, 13, 148–50, 149, 155, 169; Kinsey's findings on similarities, 77, 149; in psychology, 148–50, 168; in psychology and conditionability, 41, 157–58, 160; in responses to sexual stimuli, 158, 160, 162; in sexual behaviors, 155–57, 160
Male volume. See *Sexual Behavior in the Human Male*
Malinowski, Bronislaw, 71–72
marriage: correlation of sexual desires and behaviors in, 167–68; as difference in animal and human sexual behaviors, 151; effects of premarital petting and sex on, 114, 132, 133, 135, 152–53; importance of, 75, 81; Kinsey seeking analogous animal relations to, 153–54; sex in, 75–77, 83–84, 88, 168
marriage course, Indiana University's, 61–62, 67–86; changing content of, 68, 81–82, 85–86; complaints about, 81, 85, 192n53; evaluations of, 67, 77–78, 80–81; impact of, 80–81; importance to Kinsey's career, 64, 67, 106; Kinsey resigning from, 85, 89, 114; Kinsey's goals in, 85–86; Kinsey's lectures in, 75, 77, 81, 83–84; logistics of, 74, 79; multidisciplinary perspectives in, 80; other instructors in, 69, 74, 81, 192n53; Pearl lecturing in, 99–100, 195n31; personal conferences as part of, 78–80, 84; resources for, 69–74, 76, 189n19; Rice's, 69; students' request for, 68–69
marriage courses, at other colleges, 68–69, 78, 80
A Marriage Manual (Stone and Stone), 71
Martin, Clyde, *111*, *112*, *113*, 145; sex history interviews by, 110, 124
masculinity-femininity scale, Terman-Miles, 136, 139
Masters, William H., 6, 114
masturbation: in Davis's survey, 90–91; in humans and animals, 151; influence of class on, 127–32; in men's and women's sex lives, 148–49; as normal, 84, 85; in sex history interview, 94; treatment in other texts, 49–50; women's, 151–52, 168–69
May, Elaine Tyler, 6
Mayr, Ernst, 56, 60, 61
Meagher, John F. W., 49–50
Methods in Biology, 47–48
Mickel, Clarence Eugene, 35
Miller, Gerrit S., Jr., 72–73
modern synthesis. *See* evolutionary synthesis
Moore, W. E., 192n53
Morantz, Regina Markell, 6
morphology, taxonomy's focus on, 21–22
Mueller, John, 81
Museum of Comparative Zoology (MCZ), 33
museums, Kinsey cultivating relationships with, 31, 33

nature, 119; Kinsey's biology textbooks encouraging observation of, 19–20, 43–44, 46; Kinsey's imparting fascination with, 14, 19–20, 61
neurology, influence on sexual behavior, 160–62
nocturnal sex dreams, 148, 168–69
Nowlis, Vincent, 124

objectivity, 13, 31; in Kinsey's goals, 4, 83; nonrandom sampling decreasing likelihood of, 105–6; in sex research, 10, 83; use of orgasm as measurement and, 96–97

observation: importance in science, 4, 42, 46; Kinsey's trust in, 11, 13, 34, 65, 86, 165–66

occupational level, and social class, 126–27, 200n30

orgasm, 131; children's, 95–96, 167; clitoral vs. vaginal, 2, 155–57, 166; definitions of, 92, 96, 120; male-female similarities and differences in, 77, 148–49, 155, 157, 199n15; marriage course promoting mutual, 77, 84; sources of outlet, 120–21; use as measurement in sex research, 96–97, 120–21

The Origin of Higher Categories in Cynips, 24, 50–56

Osborn, Henry Fairfield, 35–36

outlet. *See* orgasm

pair bonding, among animals, 153
paleontology, 53
Parshley, Howard M., 72
Patten lectures, 98–100
Pauly, Philip J., 7
Pearl, Raymond, 195n31, 196n41; advocating biometry, 102–3, 122; influence on Kinsey, 93, 98–101; influence on Kinsey's statistical methods, 102–6, 122–24; on variations in behavior, 123–24

personal conferences, in marriage courses, 67, 78–80, 84

personality, Kinsey's, 18

petting/foreplay, 152, 156. *See also* premarital petting

physiology: in *Female* volume, 166; male-female similarities in, 155; in preparation for coitus, 152, 157

Pomeroy, Wardell B., 94, 95, *113*, 202n9; biography of Kinsey by, 6, 155; on *Female* volume, 143, 145; on Kinsey's gall wasps, 28–29; on sex history interviews, 78, 92; sex history interviews by, 110, 124

power: in knowledge production and transmission, 171; in sexual relations, 73, 162, 166

pregnancy, influence of fear of, 162

prejudice, 42

premarital petting, 152; influence on marriage, 133, 135; Kinsey's psychological explanations for, 148–49; marriage course lecture on, 84–85

premarital sex, 72, 88; effects on marriage, 85, 114, 132, 133; influence of class on, 127–32; Kinsey tacitly promoting, 75, 82

"The Primate Basis of Human Sexual Behavior" (Miller), 72–73

prisoners, data not included in *Female* volume, 12, 145–47

professional organizations, Kinsey's involvement with, 11, 31, 50–51, 59–61

professional relationships, 90; Kinsey cultivating, 11, 15, 21, 31, 38–39; Kinsey's vs. Wheeler's, 21; trading and giving gall wasps in, 31–33; Wheeler's, 20–21

Progressive Era, 42, 129

Promptov, Aleksandr, 51

psychology, 82; influence on sexual behavior, 126, 148; Kinsey attributing male-female sexual differences to, 13, 41; male-female differences in, 148–50, 155, 157–58, 160, 168

punched-card machines, 197n71; in analysis of sex history interview data, 12, 133; data analysis enabled by, 132, 137, 140; distinguishing Kinsey's sex research from previous, 114; information from sex interview forms entered onto cards for, 108–12, *109*; Kinsey developing uses for, 89, 107; Kinsey team acquiring, 110–11; Kinsey team using, 8, 110–14, *111, 113*; Pearl promoting, 196n41; sex interview forms relation to punched cards, 110–14

quantitative analysis, 4, 8, 12

race, 50, 198n76; in *Edible Wild Plants of Eastern North America,* 40–41; nonwhite subjects not included in Kinsey

Reports, 94, 146–47; on sex history interview forms, 41, 94
Ramsey, Glenn: development of 0–6 scale and, 85, 136–37; sex history interviews and, 92–93, 124
rape, Kinsey not addressing, 73, 88
reproduction, 58–59, 63; Kinsey not focusing on, 101, 206n76; in marriage course, 76–77
A Research in Marriage (Hamilton), 92, 120
Reumann, Miriam G., 6
Rice, Thurman B., 50, 69, 206n69
Robinson, Paul, 6–7, 159, 162, 196n53, 200n30
Rockefeller Foundation: ending funding for Kinsey's research, 167, 171; funding ISR, 110; having ASA team analyze statistics, 145
Roeth, Natalie, 25–26
Rosenzweig, Louise Ritterskamp, 21

sample size, 26, 143; benefits of huge, 34–35, 101, 103–4; insufficient to prove Kinsey's theories, 56, 103–4; Kinsey justifying choice of, 101, 118, 123–24; Kinsey's commitment to huge, 7–8, 13, 39, 103; Kinsey's compared with other sex research, 4, 119, 121; Pearl's influence on Kinsey's, 101, 103; relation to data analysis and conclusions, 4, 101, 118–19, 129; statistical methods' relation to, 101, 103–4
sampling, 101; diversity of Kinsey's, 120, 146; Kinsey comparing with other sex research literature, 121; Kinsey justifying choices in, 102, 104–6, 122, 124, 145–46; Kinsey's methods of, 97–98, 104–5; in *Male* volume, 117, 121, 124, 196n53; meaning of "normal" and, 169; nonrandom, 104, 196n50; nonwhite subjects not included in Kinsey Reports, 146–47; one hundred percent method of, 97–98, 104, 106, 119–20, 146; partial group, 98; probability, 97–98, 145–46; stratified subgroups, 104–5, 122, 124; ultimate groups sharing traits, 124, 129
Sax, Karl, 16–18

Schmitt boxes, Kinsey's gall wasp collection in, 5, 27–28
science, 8, 17, 26; applied to human issues, 21, 130; changes in, 2–5; classification in, 3, 9; creation of new knowledge in, 3, 5, 9; history of, 7–8; Kinsey in education on, 50, 64–67; Kinsey's interests in, 24, 43; Kinsey's methods in, 8, 13, 15–16; Kinsey's philosophy of, 21, 38, 61; Kinsey's skills in, 8–9, 34, 58, 61–62, 64, 86–87; Kinsey's values in, 2–3, 16; organization of data in, 5–7; taxonomy *vs.* experimentalism, 118–19; Wheeler as model for Kinsey in, 11, 15, 18–21. See also biology; entomology; life sciences; social science
Science of Human Reproduction (Parshley), 72
scientific method: applied to sexuality, 49, 68; in Kinsey's biology textbooks, 42, 44–46; Kinsey's trust in, 49, 68, 86; Kinsey teaching students, 61–62
scientists, Kinsey teaching students to be, 26–27, 61, 66
sex: Kinsey's beliefs about, 88–89; love and, 82–83; in marriage, 75–77, 83–84
sex education: beneficial effects of, 70–71, 75; effects of lacking, 135; Kinsey's beliefs about, 48–50, 75, 92; in Kinsey's biology textbooks, 42, 44, 46; Kinsey's criticism of other texts for, 49–50; Kinsey's interest in, 11, 15, 63; Kinsey teaching, 14, 65, 68–69, 84; in Kinsey's textbooks, 3, 48
sex histories: for *Female* volume, 145, 148; Kinsey moving beyond, 163
sex history interview forms, 29, 197n71; coding of, 94, 109, 109–10, 122–23; number of data points on, 114, 121–22; race on, 41; relation to punched cards, 110–14
sex history interviews, 190n32; with children, 95–96; compared with other sex researchers, 91, 121–22; development of, 11–12, 49, 63–64, 88, 93; explanation in *Male* volume, 117–20; influences on, 49, 90, 125; interviewers coding lies and hesitations in, 123;

sex history interviews (*cont.*); by Kinsey, 11, 92–93, *93,* 142; limits on Kinsey's, 170–71; number of questions, 121–22; as part of marriage course, 78–79; subjects for, 61, 79–80, 97

sex offenders: castration of, 158–59; Kinsey using information from, 88, 95–96, 167

sexology: classification as foundational tool in, 170; as interdisciplinary science, 107; Kinsey seen as founder of, 164

sex research, 10, 61, 114; benefits of, 2, 81; CRPS-funded, 90; Exner's, 97–98; for *Female* volume, 142–43; focusing on processes *vs.* individuals, 59–60; Hamilton's, 92; impact of Kinsey's, 2, 80, 169; influences on Kinsey's, 49, 88, 90, 98; Kinsey's compared to other, 89, 107, 114, 117, 119–21; Kinsey's goals in, 86, 117, 142–43; Kinsey's guiding principles for, 83, 85–86, 98; Kinsey's marriage course and, 67–68, 85; Kinsey's methods in, 15, 29, 45, 88, 110, 169; Kinsey's move to, 2, 4–5, 11, 21, 117–18; Kinsey's skills brought to, 61–62, 64, 68, 134; limits on Kinsey's, 170–71; orgasm use as measurement in, 96, 120–21

sex research literature: audiences for, 107; Kinsey reviewing, 121–22

sexual behavior, 41; analyzed by age, education, class, etc., 112–13; animal *vs.* human, 72, 98–100, 144, 150–51; of animals, 156, 158; children's, 95–96; classifications of, 134–35, 144; consent assumed, 190n32; correlation within marriage, 167–68; differences and similarities in (*See* male-female differences and similarities); diversity of, 12, 68, 77, 83, 117, 125, 162–63; in *Factors in the Sex Life of Twenty-Two Hundred Women,* 70; frequency of, 112–13; hierarchies of, 140; ill-effects of classifications of, 130, 134–36; ill-effects of sublimating, 70–71; influence of class on, 122, 125, 127–32, 146; influence of culture on, 80, 83, 161–62; influence of hormones on, 158–61; influences on, 12, 125–26, 128–30, 143, 147–48, 155–56, 160–61; Kinsey filming, 155, 164; lack of judgment of, 21, 49–50, 90–91; monitoring and control of, 128, 130, 151–52, 162; moral and religious influences on, 134–37; normality of most, 86, 100, 117, 136–37; on 0–6 scale, 136–40; as scientific interest, 17, 59, 101; stability and change in, 128, 146

Sexual Behavior in the Human Female, 82; anthropological data in, 150–52; classification practices in, 143–44; compared to *Male* volume, 145–48; comparing human and animal sexual behavior, 144, 150–51; conclusions in, 144, 161; on extramarital sex, 153–54; focusing on male-female differences and similarities, 147–48, 150, 155–57, 167; goals for, 141, 143; on influences on sexual behavior, 12, 155–56, 158–61; on monitoring and control of sexual behavior, 151–52, 162; research and writing of, 142–43; responses to, 166; selection of data for, 145, 146–47; sources for, 143, 145, 147–48, 157; statistical analysis in, 145–47, 150; 0–6 scale in, 138. *See also* Kinsey Reports

Sexual Behavior in the Human Male, 12, 117, 124–25; classifications and, 42, 116–17; data collection for, 114–15; data interpretation in, 121, 125, 132; explanation of methods in, 104–5, 118; *Female* volume compared to, 143; goals for, 122, 143; on homosexuality, 136–40; on ill-effects of classifications of sexual behavior, 134–36; Kinsey focusing on, 60–61, 66; other sex research and, 114, 119–22; publication of, 115; responses to, 116, 133–34, 142, 146, 167; sales of, 142; sources for, 140–41; statistical tables in, 103, 130–32; statistics in, 145, 167, 196n39, 196n53; 0–6 scale in, *138. See also* Kinsey Reports

sexual feelings: in Davis's survey, 91; desire, 139, 160–61, 167–68; in sex history interview, 94

sexual identity: hormones and, 136–37; as spectrum, 79, 91–92, 117, 132, 136–40

sexuality: effort to understanding totality of, 6, 8, 13, 141; Kinsey rarely using term, 148, 168–69; Kinsey's interest in studying, 14, 68, 80; Kinsey's recommended authors and resources on, 48–49; Kinsey wanting to understand totality of, 10, 163, 167, 168–69
The Sexual Lives of Savage in North-Western Melanesia (Malinowski), 71–72
sexual myths, 147
sexual positions, 133
sexual problems, 21, 78–79, 168
sexual response, psychology of, 82
sexual stimuli, male-female differences in responses to, 158, 160, 162
Shadle, Albert, 151, 154–56
Shera, Jesse H., 3, 9, 37
Simpson, George Gaylord, 53–56, 185n48
Sleigh, 20
Smocovitis, V. Betty, 58
Snedecor, George W., 196n39, 196n50; influence on Kinsey, 101–2, 104–6, 123–24
social sciences, 4, 116
Society for the Study of Evolution (SSE), 59
Society for the Study of Speciation (SSS), 50–51, 59–61
Sorokin, Pitirim, 4, 126
speciation, 4–5, 36, 51, 60
statistical analysis, 97; ASA team's critique of, 12, 145–47, 196n39; biometry, 102–3; criticism of Kinsey's, 167; in *Female* volume, 143, 145–47, 150; influences on Kinsey's, 101–2, 145–47; Kinsey comparing with other sex research literature, 121–22; Kinsey justifying, 12, 123–24; in *Male* volume, 145, 167, 196n39, 196n53; not allowing extrapolation, 103; relation to sample size, 101, 104
Statistical Methods Applied to Experiments in Agriculture and Biology (Snedecor), 101–2
statistical tables, in Kinsey Reports, 112, 117, 130–32
Stevens Institute, Kinsey at, 15
Stone, Abraham and Hannah M., 71
"Supra-Specific Variations in Nature and in Classification from the View-Point of Zoology," 53

taxonomy, 25, 35, 51; applied to human issues, 15, 117–19; Bussey Institution training students in, 16–17; criticism of Kinsey's, 166–67; experimentalism *vs.*, 118–19; genetics and, 53, 60; importance for entomologists, 23–24; Kinsey applying to sex research, 86, 114, 134; Kinsey in, 2, 3, 34, 53; Kinsey's training in, 8, 15, 21–24, 34; Kinsey's use of, 7, 68; Kinsey teaching, 61, 64–65; morphology in, 21–22; purposes of, 1, 3, 5, 15; speciation and, 23, 36; splitting and lumping in, 34–35
Taylor, William S., 48, 70–71, 74, 90
teaching standards, Indiana state, 3
technology, 114. *See also* punched-card machine
Terman-Miles masculinity-femininity scale, 136, 139
Terry, Jennifer, 6
textbooks and workbooks, Kinsey's, 3, 11, 14, 38, 42–50, 61–62. *See also specific titles*
Torre-Bueno, J. R. de la, 60
transsexual/transgender individuals, 163

Voris, Ralph, 47; Kinsey's correspondence with, 15, 36, 51–52, 68, 80

W. B. Saunders, Kinsey Reports published by, 115, 147–48
Ward, Lester Frank, 73
Warner, W. Lloyd, 4, 126, 128
Wells, Herman B, 78, 81, 85
Wheeler, William Morton, 16, 106; on biologists' perspective on human problems, 15, 18–20, 49, 68; on importance of individuals, 23–24; influence on Kinsey, 20–21, 35, 45, 66; influence on Kinsey's research methods, 15, 17–18; Kinsey's training in entomology under, 10–11; pedagogy of, 22–23; training taxonomists, 21–22, 34
World War II, 129
Wright, Clifford A., 136
Wright, Sewall, 65–66

Yerkes, Robert, 110, 169

0–6 scale, on hetero-/homosexuality: development of, 2, 85, 117; effects of, 132, 166, 169; in *Male* volume, 136–40, *138*

zoology, Kinsey teaching, 64